Data Center Critical Facilities Migration

From Construction to Operations – A Comprehensive Guide

Yekini K. Tidjani

ISBN

Hardcover: 978-1-969844-95-9

Paperback: 978-1-969844-94-2

Dedication

This book is for all the Data Center Critical Facilities Managers, IT folks, and construction crews who work hard every day to keep our digital world running smoothly. Your skills, grit, and tireless effort to keep things running smoothly are what make modern technology possible.

This book celebrates your hard work and aims to serve as a helpful guide for your ongoing success in the fast-changing world of data center management. It's also for the new generation of CFM pros, those just starting out and looking to carry forward the high standards set by those before them. I hope this book serves as a trusted friend, a guide, and a useful tool as you grow in this challenging yet rewarding field.

Lastly, this book is about the love of learning, the constant drive to improve, and adapting in this ever-evolving field. The drive to improve and innovate in data center management shows the strength and creativity of everyone who pours their heart into this essential work.

Preface

Getting a data center from the construction phase to full operational status is a huge and critical task. It requires detailed planning, tight teamwork, and a solid understanding of the complex systems at play. A smooth handover minimizes disruptions, enhances efficiency, and ensures IT services stay online without a hitch. However, this challenging process often hits snags, such as unexpected technical glitches or juggling different team members with conflicting priorities.

This book is designed to help Data Center Critical Facilities Managers and other professionals with a clear, practical guide for navigating this challenging stage. Drawing on years of hands-on experience in data center construction and operations, this handbook outlines straightforward steps, detailed checklists, and real-world examples to guide you through every step of the transition.

From setting clear goals and assembling a top-notch team to thoroughly testing systems and implementing solid procedures, this book shares smart ways to manage risks, maximize your resources, and ensure a seamless transition. Beyond the technical aspects, we delve into the vital roles of communication, teamwork and collaboration with others. Getting a data center up and running smoothly requires more than just technical expertise—it also demands strong people skills and a team-first attitude. This book is designed to equip you with the tools and knowledge to excel in both technical and interpersonal aspects, paving the way for a successful data center transition.

The guidance here is intended to be a comprehensive resource for both seasoned professionals seeking to refine their approach and newcomers eager to establish a solid foundation in this field. We hope this book empowers you to lead your team with confidence, creating a reliable and efficient tech backbone that supports everyone who depends on it. With these strategies, you can tackle challenges directly, resolve problems effectively, and keep your data center running smoothly, delivering the stability and performance that today's tech-driven world demands.

Table of Contents

Introduction

The modern world runs on data. The critical facilities that house and protect this data—our data centers—are the unsung heroes of our digital age. Their flawless operation is essential for businesses, governments, and individuals alike. Yet, the journey from the blueprint to fully operational status is far from simple.

The transition phase between construction and operations presents a unique set of challenges that require careful planning, skilled execution, and unwavering attention to detail. This book is your comprehensive guide to navigating this crucial period. It offers a practical, step-by-step approach to data center transitioning, addressing every aspect from initial planning to ongoing maintenance.

We begin by addressing the essential aspects of pre-transition planning, emphasizing the need to set clear objectives, assemble a competent team, and conduct thorough risk assessments. Subsequent chapters cover system testing and commissioning— a critical yet often overlooked or poorly managed phase. Here, we provide a detailed framework to ensure that all systems are fully functional and meet performance requirements before handing them over to the operations team. Staff training and knowledge transfer are also integral components of a successful transition.

We explore strategies for developing effective training programs, fostering a culture of continuous learning, and establishing a robust knowledge base for the operations team.

Later chapters guide you through the development of comprehensive operational procedures, the management of electrical power and mechanical cooling systems, addressing networking and security concerns, effective strategies for managing tenants and their diverse needs, and the importance of establishing robust work approval processes and effective governance.

We also highlight how to leverage tools such as Data Center Infrastructure Management (DCIM) platforms, Building Management Systems (BMS), and collaboration tools like SharePoint to enhance visibility and control. Ultimately, we emphasize the value of post-transition reviews as a catalyst for ongoing improvement.

This book is written for Critical Facilities Managers, IT managers, commissioning agents, and professionals involved in the construction and transition of data centers. Whether you're a seasoned expert or new to the field, this book equips you with actionable strategies and proven practices to lead successful transitions and ensure operational excellence.

Establishing Clear Objectives and CPIs

Setting clear goals and Key Performance Indicators (KPIs) and Critical Performance Indicators (CPIs) is fundamental to a successful data center transition. Without them, the process can easily lose direction, become inefficient, and risk an unstable handover from construction to full operation. This section offers a practical framework for defining objectives and aligning Key Performance Indicators (KPIs) and Critical Performance Indicators (CPIs) with broader business goals, ultimately building a focused, measurable, and actionable transition plan.

- The first step is figuring out the business goals behind the data center project. Is it driven by the need to expand capacity, enter a new market, support a new business function, or reduce operational costs? These big-picture goals shape the specific aims for the transition phase. For instance:
- If cost reduction is the main driver, transition efforts should emphasize efficiency and resource optimization.

If uptime is a critical concern, focus shifts to system resilience and operational continuity

Once the big business goals are clear, set SMART goals—Specific, Measurable, Achievable, Relevant, and Time-bound—across critical areas such as power, cooling, networking, security, and tenant management. Here are some examples:

Uptime: A key Performance Indicators (KPIs) and Critical Performance Indicators (CPIs) is system uptime. Set a clear target, such as achieving 99.99% uptime within the first 90 days post-transition. This requires a robust monitoring system from the outset and a well-defined plan for addressing problems. The transition should include thorough testing of all systems, including backups, to ensure the target is achievable. If the target isn't met, review the plan and find ways to improve.

Cost Efficiency: Keeping track of spending—both capital expenditures (CAPEX) and operational expenditures (OPEX)—is vital. The transition plan should have a detailed budget and closely monitor costs. Key Performance Indicators (KPIs) and Critical Performance Indicators (CPIs) may include staying within a set budget range, achieving a specific Return On Investment (ROI) within a certain timeframe, or reducing energy consumption per computing unit. This requires careful planning and resource management, often involving the use of project software to track and report costs. Regular budget checks, conducted perhaps monthly, can identify potential overruns and enable quick fixes.

Tenant Satisfaction: For data centers serving multiple tenants, maintaining tenant satisfaction is crucial. The transition plan should include regular communication with tenants, addressing their concerns, and sharing updates on progress. Key Performance Indicators (KPIs) and Critical Performance Indicators (CPIs) could include tenant satisfaction surveys, the speed of respond to tenant requests, and the number of Service-Level Agreement (SLA) breaches. Clear communication, solid SLAs setting expectations, and a clear process for handling issues are key to keeping tenants satisfied.

System Performance: The transition plan should establish clear measures for evaluating system performance, including network speed, storage response times, and server utilization. These measures need to be closely monitored after the transition to spot any slowdowns or issues. Regular testing, both during and after the transition, will help confirm that the system meets the set goals and identify where improvements are needed.

Security Compliance: Security is a top priority in data centers. Key Performance Indicators (KPIs) and Critical Performance Indicators (CPIs) should track key metrics, such as the number of security issues, the speed of resolution, and whether the center adheres to industry regulations and standards. The transition plan should include detailed security checks and tests to find and fix weak spots. Meeting security standards, such as SOC 2, ISO 27001, or

PCI DSS (depending on the data center's purpose), should be carefully recorded and tracked.

Besides setting specific Key Performance Indicators (KPIs) and Critical Performance Indicators (CPIs), the transition plan needs a clear timeline with milestones and deadlines for each step. This timeline should be practical, allowing for possible delays or surprises. Tools like Gantt charts or project management software can help visualize the timeline, highlight task dependencies, and track progress.

The transition plan should also include a robust communication setup to ensure everyone is on the same page. This setup should outline communication methods, the frequency of updates and reporting steps. Regular meetings, utilizing tools like Microsoft Teams or similar platforms, can facilitate smoother teamwork and communication among all the teams involved.

Aligning transition goals with the company's broader objectives is crucial. This ensures the data center transition supports the organization's overall plans. By setting clear goals and carefully tracking Key Performance Indicators (KPIs) and Critical Performance Indicators (CPIs), the Critical Facilities Manager can ensure a smooth transition, keeping disruptions low and efficiency high.

This also establishes a system for ongoing performance checks and steady improvements following the transition. Regularly reviewing the plan and learning from what worked or didn't will help adapt to new needs and ensure the data center's long-term success. The recorded process, including challenges and fixes, will serve as a valuable resource for future projects.

Ultimately, delivering a successful transition, backed by performance data, strengthens the credibility of the Critical Facilities Manager and their team. It demonstrates capability, builds stakeholder trust, and positions the team to lead future complex initiatives with confidence.

Assembling the Transition Team and Defining Roles

Assembling a competent and cohesive transition team is the cornerstone of a successful data center handover. This team requires a diverse set of skills, encompassing construction, operations, IT, and commissioning. Each person brings something special to the table, and picking the right people with clear roles helps avoid conflicts and keeps things running smoothly. The team's size and structure should align with the project's scope. A smaller, nimble team might be suitable for a simple project, while a larger, more organized team is needed for large, complex data centers.

Begin by selecting key individuals from each department. For the construction team, appoint someone thoroughly familiar with the facility's final design, system configurations, and any outstanding tasks. This person will be essential in answering questions and ensuring the transition runs smoothly." Their primary responsibility is ensuring that the building and systems meet operational requirements and comply with all applicable rules and codes. They also serve as the primary point of contact for any ongoing construction work as operations commence. Select experienced professionals from the operations team who have a deep understanding of critical data center systems, such as power, cooling, and security. Their expertise will be crucial for establishing effective operational procedures and addressing issues as they arise during the transition. Their responsibilities include developing work procedures, thoroughly testing systems, and training the operational team. They ensure everything runs smoothly and meets the set performance goals. The IT team plays a crucial role in verifying the network's readiness, testing storage systems, and ensuring that new systems integrate seamlessly with the existing infrastructure. The IT expert should be well-versed in server configurations, network management, and application deployment. Their role extends to confirming that the IT infrastructure can handle operational demands and assisting with the migration of applications and data.

Additionally, they are responsible for setting up monitoring tools to track performance and identify potential IT issues early. The commissioning team, comprising expert engineers, ensures the facility operates according to its design specifications. They conduct extensive testing and document systems performance. Their reports confirm that the data center meets all required standards and support warranty periods. As the final phase of the transition approaches, they verify system performance and ensure a smooth handoff to operations. Their sharp eye for spotting potential problems helps fix issues before they grow. A strong team like this build's confidence, keeps the transition on track, and sets the data center up for long-term success by tackling challenges early and working together seamlessly.

To ensure smooth coordination, setting clear roles and responsibilities for the transition team is a must. A detailed chart spelling out who is responsible for what keeps things organized and avoids confusion or gaps. This chart should be checked and updated regularly to accommodate changing needs and team shifts. Holding regular team meetings and utilizing tools like Microsoft Teams or SharePoint is crucial for maintaining open communication, resolving issues, and keeping the project on track. These meetings provide everyone with an opportunity to address problems promptly and prevent minor disagreements from escalating into major issues.

Effective communication is the heart of a successful transition. This involves establishing clear channels for sharing information, using straightforward and consistent terminology, and maintaining a comprehensive record of all communications. A central platform like SharePoint is a great tool for easily sharing documents, updates, and progress reports. Regular meetings and transparent communication help mitigate risks, resolve conflicts, and keep everyone on the same page throughout the transition. When roles and communication are unclear, the consequences can be severe. Imagine the construction team wrapping up without telling the operations team about small but important details. This could lead to downtime, inefficient operations, or even safety risks. For example, if critical shut-off valves are not located due to missing paperwork, it could slow down emergency responses and damage equipment. Similarly, if the IT team fails to thoroughly test network connections before the final switch, it could lead to long downtimes and upset tenants. The result? Delays in getting the data center ready, frustrated tenants, and possibly hefty fines for missing service agreements.

That's why having ways to sort out conflicts is so important. Everyone should feel okay speaking up about concerns or disagreements. A clear process for handling conflicts, perhaps with a senior manager serving as a fair mediator, should be in place to prevent things from escalating. Team-building activities can help create a positive atmosphere, fostering trust and respect among team members, which in turn enhances communication and teamwork.

A strong transition team requires a capable leader, someone with experience in managing data center transitions and a proven ability to work with diverse teams. This leader oversees all aspects of the project, manages risks, resolves conflicts, and facilitates effective communication with stakeholders. Their role involves providing regular updates to senior management, removing obstacles that hinder progress, and maintaining the team's motivation and focus on the overall objective. The transition leader must have excellent organizational, communication, and problem-solving skills to guide the team toward success.

A good leader keeps the team focused, ensures everyone's working together, and builds confidence that the transition will succeed, setting the data center up for long-term success. The transition team's success depends on staying ahead of risks. Spotting potential problems early, determining their likelihood and potential impact, and devising effective solutions to address them is crucial. A detailed risk assessment should be conducted at the outset, and the plan requires regular updates as the transition progresses. This check should examine a range of issues, including equipment breakdowns, power failures, security problems, and errors by team members.

Having clear backup plans for each risk is vital to ensure the team can act quickly and keep things on track if something unexpected happens.

In addition to their specific roles, each team member must receive thorough training on data center operations, security protocols, and emergency procedures. This training ensures that everyone is prepared to handle potential challenges during the transition. It should include practical instruction on using monitoring systems, such as EPMS and BAS, troubleshooting issues, and responding to incidents. Training should be tailored to each person's role, ensuring that everyone has the necessary skills and knowledge to perform their job effectively. Regular refresher courses should keep the team up to date on new technologies and industry best practices. Keeping comprehensive records of training ensures the team meets industry standards and remains well-prepared for the challenges ahead. Finally, keeping good records throughout the transition is crucial. This includes detailed notes from meetings, progress updates, risk assessments, and other key information. These records act as a history of the transition, offering lessons for future projects. They also support decisions, keeping things transparent and accountable. A detailed checklist of tasks ensures nothing important gets missed. These records should be easily accessible to the team and stakeholders through a central project platform. This clear record-keeping will be a huge help in future audits, making it easier to demonstrate that everything was done correctly and supporting any warranty claims.

Comprehensive documentation, including insights gained from the transition process, will help the Critical Facilities Manager and their team improve future project planning. It strengthens the team's ability to address challenges and lays a solid foundation for running the data center efficiently long after the transition is complete.

Risk Assessment and Mitigation Strategies

The pre-transition phase is a critical period that requires careful planning and proactive risk management to ensure a smooth handover to operations. A thorough risk assessment is not just a formality—it is a vital step to protect the entire data center transition. Failing to conduct one can lead to expensive delays, operational hiccups, or even major failures. Below, we'll walk through a clear method to identify, analyze, and tackle these risks.

Begin by establishing a risk register. This is more than just a basic list—it should be a living document, regularly updated as the transition progresses. For each risk, provide a clear description, asses how bad it could get (from a small hiccup to a full-blown disaster), and evaluate the likelihood of it happening (from "probably won't" to "almost certain"). A simple risk matrix works well here; plotting likelihood against impact helps determine which risks require immediate attention.

High-Likelihood, High-Impact Risks need to be addressed first. Let's examine some common risks during data center transitions:

Equipment Failures: These can range from minor glitches in UPS systems to complete breakdowns of the cooling system. The fallout could be anything from a brief outage to a complete shutdown. To avoid this, test all equipment rigorously before going live, build in backups like N+1 redundancy for power and cooling, and keep spare parts on hand. Good relationships with vendors are key, and setting clear SLAs in place guarantees fast repairs. Your risk register should list every critical piece of equipment, along with vendor contacts, promised response times, and the location of backups.

Construction Delays: Unexpected delays can derail the timeline. Issues like material shortages, site problems, or subcontractor delays are common culprits. To stay ahead, set realistic schedules with sufficient buffer time, closely monitor progress, and maintain constant communication with the construction team. A solid change management process helps handle surprises smoothly. Also, have backup plans in place, such as extending the timelines or arranging temporary spaces. Document potential delay causes in your risk register, including their impact and damage-control strategies.

Vendor Coordination Issues: Data center transitions often rely on multiple vendors delivering and installing equipment. Poor coordination can mean missed deadlines and chaos. To prevent this, establish clear lines of communication, create a detailed installation schedule, and implement a process for handling changes. Regular check-ins with vendors help catch problems early. In your risk register, note key vendors, their roles, potential risks, and backup plans (e.g., alternate suppliers or workarounds in case a vendor fails to fulfill their obligations).

IT System Integration Issues: Integrating new IT systems with existing setups can become complicated quickly. If not done properly, you're looking at long downtimes and lost data. To prevent this, test each system thoroughly, independently and together simulating real operations before switching over. Have backup plans ready in case things fail, and consider bringing in experienced IT staff to handle the merge. Your risk list needs step-by-step checklists and tests, plus every possible weak spot in your IT setup flagged upfront.

Security Breaches: During transitions, security risks increase due to the presence of extra construction crews and temporary access points. Secure your premises with strict access rules, limit entry to authorized personnel, and train all staff on security basics. Keep a constant eye on security 24/7—intrusion alarms are a must. Your risk register should track all security steps, any weak points you've identified, and exactly who is responsible for responding if something happens, including who to contact in the event of a serious issue.

Power Outages: Power failures, whether planned or unplanned, can cripple a data center. Protect yourself with backup power sources like UPS batteries and generators), solid monitoring tools, and clear emergency steps. Test those backups regularly. The risk register should outline what your backups can handle, including their duration, fuel supplies, and who to notify (both tenants and power companies) in the event of trouble.

Cooling System Failures: A failure in the cooling system can result in equipment overheating and data loss. Prevent this by installing backup cooling units, performing routine system checks, and ensuring systems are monitored. Test the cooling systems at full capacity before launch and document in the risk register.

Human Mistakes: People screw up—it's why data centers have problems. Fight this with proper training, practice drills, crystal-clear instructions, and regular check-ins. Your risk register should identify common slip-ups and guide how to avoid them, along with simple checklists and safety rules. Good planning means more than just spotting risks—you need action plans. For each danger, map out exactly what to do if it happens. Revisit these plans frequently, as circumstances may change. Keeping your risk register updated and discussing it in team meetings means fewer unpleasant surprises. This organized approach to handling risks makes your data center migration safer and smoother.

Your risk register is not just another document—it serves as proof of your due diligence and provides valuable insights for future projects. By regularly reviewing and updating it, your team remains vigilant about potential risks and ensures a proactive approach to resolving them.

Developing a Comprehensive Transition Plan

Building on the solid risk assessment we've already completed, the next major step is to put together a comprehensive transition plan. This isn't just some basic checklist—it's a flexible guide that will take us from finished construction to full operation. When we nail this plan down and follow it carefully, we'll ensure a smooth handoff, minimize problems, and maximize the benefits of our operations. We must treat this plan as if it were alive, checking and updating it regularly as we move forward and learn new things.

What makes this transition plan work? Crystal clear job assignments. Every team—construction, operations, IT, and commissioning—must be aware of their specific responsibilities and deadlines. Clear roles help us avoid confusion, missed deadlines, and interpersonal conflicts. A Responsibility Assignment Matrix (RAM) is a crucial tool in this context. It's a chart that assigns each task to a specific individual or team, ensuring accountability and transparency. Each phase should have clear milestones that mark important steps toward full operation. These checkpoints allow us to see how we're doing and catch any issues early. Using a Gantt chart, such as the one in Microsoft Project, provides a clear picture of our timeline, how tasks connect, and what is most critical. Seeing it all laid out helps us spot potential conflicts and adjust the schedule as needed.

A detailed transition plan should address critical aspects like these:

System Testing: Testing isn't just a one-time activity; it's a step-by-step process. Start with testing individual components, then move to Integrated System Testing (IST), and finally conduct full-load testing under conditions similar to real-life operations. Each testing phase should follow an approved IST script with clear procedures, defined outcomes, and acceptance criteria. If issues arise, they should be logged with precise details, including what went wrong and how it was resolved. After testing, detailed reports should be shared with all stakeholders on a central platform for easy access.

Staff Training: Before the official handover, the operations team must undergo thorough training. This training should cover operating procedures, troubleshooting common problems, and utilizing systems such as electrical power monitoring and building automation tools. Tailored sessions ensure each team member gains the skills needed for their specific role. Post-training assessments will identify any knowledge gaps, allowing us to address them before the full transition.

Documentation: Maintaining well-organized records is crucial for a smooth transition. This includes as-built drawings, equipment specs, Methods Of Procedure (MOP) steps, and maintenance guidelines. All documents should be reviewed and approved of by relevant stakeholders before the handover and stored securely for easy access. Emphasis should be placed on version control and ensuring document security.

Vendor Management: The transition plan must clearly define vendor roles, including ongoing maintenance contracts, Service-Level Agreements (SLAs), and emergency support. Establishing effective communication channels and clear escalation paths is crucial to resolving issues promptly. Regular communication with vendors— both before and after handover—will ensure strong working relationships. All agreements and contact details of key vendor representatives should be documented for easy reference and future reference.

Emergency Response: The plan must include detailed emergency protocols for handling power outages, cooling system failures, or security issues. These should clearly assign roles, outline communication steps, and explain escalation processes to ensure safety and quick service restoration. Regular drills will help staff practice these procedures and fine-tune them for efficiency. Feedback from these drills should be documented to improve future processes.

Tenant Coordination: If the data center houses multiple tenants, the plan should outline communication and coordination strategies to ensure a seamless transition without service disruptions. A clear communication plan should be established, explaining the transition process, timelines, and potential impacts to the tenants. This proactive approach will mitigate potential conflicts or misunderstandings.

Let's look at a transition plan for a medium-sized data center, broken down into four key phases:

Phase 1:

Pre-Commissioning (4 weeks): This phase focuses on testing individual components to ensure they function as expected and meet the specified requirements. Key tasks include testing each piece of equipment, verifying the runtime of UPS and generators, and thoroughly evaluating the cooling systems. Clear responsibilities are assigned to the commissioning team, with regular updates shared with the construction and operations teams to ensure alignment and seamless integration.

Phase 2:

Integrated System Testing (2 weeks): This phase evaluates the integration and functionality of all systems under operational conditions. Milestones include completing integrated tests, confirming that redundancy measures are functional, and demonstrating the effectiveness of disaster recovery plans. During this phase, the commissioning team, general contractors, subcontractors, and control teams work closely together. They log any issues that arise and implement the necessary fixes to ensure everything runs smoothly.

Phase 3:

Staff Training and Documentation Review (2 weeks): During this phase, the operations team undergoes training, and all supporting documents are reviewed and finalized. Key tasks include completing the staff training program and obtaining final approval for as-built drawings and operational procedures.

Phase 4:

Handover and Go-Live (1 week): This is the final step, where the data center becomes fully operational. Tasks include handing over responsibilities to the operations team, signing off on commissioning, and officially starting operations. All teams work together to ensure this phase is executed without any major interruptions.

To support all phases, project management tools like SharePoint are essential. SharePoint helps track progress, organize tasks, and centralize communication and documentation. Real-time updates enable easier identification and early addressing of potential problems. By storing documents centrally, the platform allows every team member to access important information quickly. It also provides a way to record any changes to the plan, ensuring it remains accurate and up-to-date. Regular reports generated through SharePoint keep stakeholders informed and support effective decision-making throughout the process.

Creating a comprehensive transition plan, with these elements in place and leveraging tools such as SharePoint, is crucial for success. Regular updates, teamwork, and attention to detail help ensure a seamless transition. The result is a fully operational data center that meets all expectations and operates efficiently. Careful planning in this stage sets the foundation for smooth handovers and long-term success.

Budgeting and Resource Allocation

Budgeting and resource allocation are essential for a successful data center transition. Without proper planning for financial and human resources, projects can face delays, quality problems, or even outright failure. A detailed budget, developed in conjunction with the transition plan, helps mitigate these risks and ensures the project runs smoothly.

The budget should include all anticipated costs, such as spare parts, material purchases, software licenses, and a reserve for unexpected expenses. Start by categorizing costs into areas such as corrective and preventive maintenance. Carefully review the plan and list every task, along with its resource needs, including salaries for all team members—contracted security personnel, temporary staff, managers, engineers, and technicians. Construction costs should also cover overtime, travel, and training. Material costs should include the tools, supplies, and consumables needed for testing, maintenance, and operations. Examples include electrical equipment, mechanical parts, and cable management materials. Licenses for software like project management tools, DCIM systems, and building management software should also be factored in. Beyond these core expenses, a contingency fund—typically a percentage of the total budget—should be set aside to cover unexpected costs or delays. Reviewing budgets from previous data center transitions can help determine a reasonable reserve.

Let's consider a practical example for a medium-sized data center transition:

- Personnel:

Allocate $500,000 for project managers, engineers, technicians, and managers. Break down costs by role, such as a senior engineer at $8,000 per month for 12 months. This detailed breakdown makes it easier to track expenses, including overtime for unforeseen circumstances.

- Materials:

Budget $200,000 for maintenance items like electrical and mechanical equipment, cable supplies, and consumables. It should also cover minor hardware repairs identified during the testing process. Prepare a detailed list with part numbers and quantities, and obtain quotes from multiple vendors to ensure cost-effectiveness. Flexibility should be built into address changes in material needs during the transition.

- Software:

Budget $20,000 for licenses related to tools for managing access systems, EPMS, and BAS software, especially when redundancy is limited. This cost should also include any necessary staff training for using these tools effectively. Don't forget to account for maintenance and support costs associated with the software.

- Contingency:

Allocate $15,000 (10% reserve) for unexpected costs, such as equipment failures or project delays. This fund should be easily accessible and carefully tracked, with regular reviews to ensure it's adequate for potential issues.

- Third-Party Services:

Budget $15,000 for external consultants or contractors needed for tasks like obtaining permits or inspections. Services and costs should be clearly defined, and contracts should be carefully reviewed to identify any risks or unexpected expenses.

Total Estimated Budget: $750,000

This simplified overview of the budget provides a basic breakdown of the budget. Real-world budgets require detailed line items with justifications and supporting documentation. A transparent budget with detailed tracking creates accountability and allows for precise financial monitoring throughout the project. Make sure to include the contingency fund and implement robust tracking methods—these elements help keep the project on course and ensure the successful delivery of the data center. Adding flexibility and depth to the budget helps the team handle unexpected challenges while keeping costs under control, ensuring a smooth transition.

Resource allocation, closely tied to the budget, ensures that there are sufficient staff and materials available at each stage of the project. This requires a phased approach, aligning resources with the project timeline. In the early stages, which focus on testing and commissioning, most resources are typically assigned to the engineering and commissioning teams. Later, as the project shifts toward operations and staff training, those resources should be redirected accordingly. A clear resource allocation plan, supported by tools like SharePoint, can streamline this process, preventing delays or bottlenecks.

Monitoring expenses carefully is critical during every stage of the project. Comparing actual spending to the approved budget regularly identifies potential overspending early, allowing the team to act promptly before it becomes a more significant issue. This requires a system to record all expenses, categorize them by budget items, and compare them to forecasts. Any significant differences between planned and actual spending should be investigated immediately.

This may involve reevaluating task timelines, renegotiating contracts with vendors, or exploring new ways to reduce costs.

Regular financial reports generated using a project management system should be shared with stakeholders to ensure transparency and support informed decisions.

Managing financial risks is a crucial aspect of the budgeting process. To mitigate risks, teams should work with multiple suppliers to avoid excessive reliance on any single one, implement robust contract management practices, and maintain a contingency fund as previously discussed. Financial risks should be reviewed regularly to identify potential issues and outline strategies for their resolution. These reviews shouldn't be treated as a one-time task but as ongoing evaluations, integrated into regular project review meetings.

A realistic budget, a clear resource allocation plan, and effective financial risk management are essential for a successful data center transition. Detailed planning, accurate tracking, and proactive cost management ensure the project stays on schedule, within budget, and meets operational expectations. A well-thought-out approach to budgeting and resource allocation not only strengthens financial stability but also improves efficiency, minimizes disruptions, and ensures the overall success of the transition.

Comprehensive Integrated System Testing Procedures

The smooth transition of a data center from construction to operation depends heavily on thorough system testing. This step isn't just a routine task; it's a critical process that confirms all systems work as intended, meet their specifications, and operate together without issues. A well-planned and carefully executed testing strategy reduces operational disruptions and enhances the facility's long-term efficiency. It includes detailed planning, execution, proper documentation, and resolving any problems discovered during testing.

The first step is creating a comprehensive, integrated system testing (IST) script. This script serves as a comprehensive guide for the entire testing process, detailing the tests to be conducted, the methods to be employed, the criteria for success, and the individuals responsible for each task. The script should cover all essential systems in the data center, including power distribution (UPS, generators, and PDUs), cooling infrastructure (CRACs, CRAHs, FWUs, and chillers), network equipment (switches, routers, and firewalls), security measures (access controls and surveillance), and building management systems (such as temperature, humidity, and airflow controls). Each system needs a testing strategy tailored to its specific role and possible failure points.

For instance, testing the power distribution system may involve load bank testing to confirm UPS capacity and runtime, as well as generator performance tests under various load conditions, ranging from 25% to full capacity. Cabling and connections should also be inspected with tools like infrared scanners. The process should include measuring voltage, current, and power factors at various points in the system. These measurements ensure the system complies with industry standards and reveal any weaknesses or inconsistencies. Load bank testing should be incremental, gradually increasing the load to check whether the UPS and generator can handle peak demands. Keeping detailed records of the tests, including load levels, voltage readings, and any unusual findings, is crucial for troubleshooting and future reference. This rigorous testing process helps identify and resolve potential issues related to power stability or outages.

Similarly, testing the cooling system involves a comprehensive approach. This includes evaluating the capacity of FWUs, CRACs, and CRAHs under various load conditions, verifying the efficiency of the cooling distribution system, and confirming that the chillers are operating properly. This process should involve measuring temperatures at multiple points in the data center, including the hot and cold aisles, to ensure they stay within the acceptable range. Leak detection tests are also essential for identifying potential leaks in the chilled water system, which could lead to costly repairs or interruptions. The testing protocol must carefully document readings for temperature, humidity, and airflow across the facility to maintain consistent climate control, preventing equipment overheating and failures.

Testing the network requires a structured verification process. It begins by ensuring basic network connectivity, confirming that switches, routers, and firewalls are correctly configured and functioning as expected. Once basic connectivity is established, the process proceeds to advanced testing, including validating bandwidth capacity and measuring latency to assess network performance. Network security testing is equally important and should focus on identifying vulnerabilities that could lead to potential attacks and breaches. Tools and techniques must be used to uncover weak points and implement measures to address them. This thorough approach ensures the network remains reliable, secure, and capable of handling high-performance demands.

Security system testing is a top priority. This involves thoroughly evaluating all security measures, including access control systems, surveillance equipment, and intrusion detection mechanisms. Testing access controls involves checking that keycard readers, door locks, and similar systems are functioning correctly. Surveillance testing confirms that cameras, recording devices, and monitoring systems provide complete coverage of all critical areas. Intrusion detection systems must also be thoroughly tested to ensure they can effectively detect and respond to unauthorized access attempts. This careful and detailed testing process safeguards the data center and its critical assets.

The testing process should follow well-known methods and tools used across the industry. This helps maintain consistency and accuracy, making it easier to compare results with common standards. Using the right testing tools and software is often necessary to obtain accurate readings and reliable data. It's also essential to regularly check, maintain, and calibrate the equipment so the results stay reliable and meaningful.

During the testing stage, clear and meticulous record-keeping is essential. Every test result—whether normal or indicating problems—should be fully documented and kept in one central location. This helps keep track of everything and demonstrates responsibility for resolving any issues that arise. These records should follow a simple, consistent format. This includes the test details, the method of conducting the test, the results obtained, and the steps taken to address any issues that arose. Storing these documents digitally on platforms like SharePoint makes it easier for everyone to access and collaborate on them. Dealing with problems as they appear is very important. There should be a clear system in place for reporting, checking, and fixing any trouble found during testing. Everyone involved—construction, operations, and commissioning teams—needs to communicate clearly and openly with one another.

The most serious issues should always be handled first. A proper tracking tool, such as specialized software designed for logging issues, should be used to monitor progress and ensure that nothing is missed. This ensures that problems are fixed in a timely manner.

Once testing is done, a formal sign-off document should be created. This indicates that all the main systems have been thoroughly tested, passed the necessary checks, and are now ready for real-world use. The document needs signatures from the construction team, the design team, operations, commissioning staff, and any vendors involved. This sign-off marks the end of a key step and starts the handover to full operation. With this clear and careful process, the data center is set up for stable, smooth running, cutting down future issues, and keeping things online as much as possible.

Addressing and Resolving System Anomalies

Finding and fixing system issues during testing is a crucial part of getting a data center up and running correctly. Simply reacting to problems as they show up isn't the best approach—it wastes time and can lead to unexpected downtime. A more effective approach is to plan, communicate clearly, and employ a systematic method for problem-solving. This smarter approach doesn't just fix what's wrong right away; it also examines what caused it in the first place to help prevent it from happening again.

The first step in resolving a problem is to identify it accurately and document it in detail. That means explaining the issue clearly—what system is affected, what symptoms you're seeing, when it started, and any conditions surrounding it, such as temperature or power changes. Photos and videos, when available, are also helpful. This information helps teams resolve the issue more quickly and ensures everyone has a shared understanding. Using a shared digital system, such as SharePoint, makes it simple for different teams—construction, operations, controls, and commissioning—to stay on the same page. It should also allow people to track different versions, so no one gets confused by outdated updates.

Once the issue is clearly written up, the next step is to check for basic problems. These can include issues such as a loose wire, incorrect settings, or a malfunctioning component. This first check often catches the easy stuff and saves time by avoiding bigger delays or unnecessary panic.

However, if this quick check doesn't resolve the issue or identify the problem, a more thorough examination is necessary. A careful and step-by-step approach is best for tackling more complex problems. This involves reviewing system logs, examining performance metrics, and conducting additional tests to narrow down the issue. Each attempt that's made should be documented, including what was done, what happened, and any decisions made based on the results. Having this full history is helpful if the issue shows up again or if other people need to review what was done.

Getting to the root cause of a problem gives teams better control and teaches them something useful for next time. It also highlights potential weak points in the setup and helps prevent the same kind of trouble from recurring. These lessons are worth a great deal over time and help the entire system run better and more smoothly in the long run.

Identifying the root cause of a problem is one of the most crucial steps in resolving issues during testing. Root Cause Analysis (RCA) isn't just about solving what's wrong at the moment—it's about finding out what caused the issue in the first place. There are various methods for accomplishing this, including the "5 Whys" method, fault tree analysis, or using a fishbone diagram. The method you use should match the complexity of the issue and the tools or time available. The goal is not just to patch things up but to understand what went wrong and why. That knowledge helps teams prevent the same thing from happening again, saving time and avoiding future trouble.

Take this as an example: while doing load bank testing, the team notices odd voltage readings from one specific Power Distribution Unit (PDU). The first step is to inspect the wires and connections to ensure that everything is secure and undamaged. If nothing appears off, it means further investigation is needed. At this point, you may want to refer to the electrical diagrams, take new voltage readings from various points in the circuit, and verify the power levels on the UPS and generator.

If that doesn't yield any results, the team should test the other PDUs to determine if the problem is specific to that one or if it's a broader system issue. Eventually, a closer look inside the UPS may reveal a part that's not functioning properly. If that's the case, replacing the part and updating the maintenance plan helps ensure the same issue doesn't happen again later on.

Clear communication is just as important as the technical work. Everyone involved—the construction crew, operations team, controls experts, commissioning group, and any vendors—needs to be kept informed. Providing regular updates helps avoid confusion and ensures everyone understands where things stand. This is especially important when a serious problem might delay the project. A clear plan should be established early in the process that outlines how often updates will be provided, who will receive them, and what content will be included. This helps keep the project on track and avoids last-minute surprises. Open and honest communication fosters trust across teams and helps ensure the job is done right the first time.

Fixing problems the right way is a team effort. Everyone brings something important to the table. The construction team is familiar with the details of how the systems were assembled. The operations team understands how everything is supposed to function on a day-to-day basis.

The controls and IT teams bring deep knowledge of the network, servers, and building systems. The commissioning team focuses on testing and ensuring everything works as it should. When these teams work closely together, the chances of finding and fixing the issue correctly increase. Regular check-ins and group problem-solving help ensure that the best ideas from each team are utilized to move things forward.

Each issue that is solved should be documented clearly. That includes the steps taken, the results obtained from those steps, and the fix that was implemented. These records are more than just paperwork—they help future projects, show what worked, and point out what didn't. They also give teams the chance to review the entire process and identify areas for improvement next time. Writing things down steadily and reliably is important. It doesn't matter if the problem was big or small—each one should be handled with the same level of care. These records help everyone understand how the systems work and ensure that things run smoothly over time.

Once a problem is resolved, a formal closure report should be generated. This report wraps up the full story of the issue. It should include what the problem was, how it was addressed, what caused it, and what solution was implemented. It should also demonstrate that the fix was successful and that the system is now functioning as intended. Everyone involved should review and sign off on this report so there's a clear record of what happened.

These reports establish a shared knowledge base that facilitates smoother future maintenance and operations. Taking the time to document everything properly helps avoid repeating the same problems. The goal is not just to fix what's wrong, but to learn from it, make improvements, and ensure the data center continues to run smoothly in the long term. Teams that do this well set themselves up for lasting success and fewer surprises down the line.

Commissioning Documentation and Handover

Keeping detailed records during and after commissioning plays a crucial role in ensuring the data center transitions smoothly into full operation. These records become the go-to source for the operations team and also aid in later repairs, upgrades, or changes. Every document needs to be easy to find, simple to understand, and follow a consistent layout so that anyone who needs the info can use it without confusion. To manage everything properly, you need a clear system in place—typically on a digital platform like SharePoint—that helps track different versions and allows everyone to access the latest updates. The goal isn't just to collect random files but to create a complete, connected set of documents that work together and make cross-checking simple.

This comprehensive set of documents should encompass several key areas. First, it needs to list all the testing that got done. That includes what each test was for, the tools used, the results obtained, and any issues that arose. These details matter a great deal down the line when the team needs to make a fix. Clear test records help the operations team figure out where to start and what to check first. The documents should also include visual references, such as drawings, schematics, and diagrams, that illustrate the setup of everything—details like how the power is routed, where cooling systems are located, how network cables are laid out, and more. These images help the team visualize the full layout and identify any physical issues that arise.

Additionally, the documents must include a comprehensive list of all the equipment that was installed. This list should include the brand, model, serial number, and the frequency of regular maintenance required. Keeping this information in one place helps teams track the lifespan of each item, determine when to service it, and plan for replacements accordingly. It also helps the team understand what each piece of equipment can do and its limitations.

Knowing the maintenance schedule enables the team to plan effectively, avoid surprises, and minimize the likelihood of issues arising. It also helps avoid last-minute scrambles. Having a complete list of vendor contacts in one place means the team can get help quickly when something breaks or when parts are needed urgently. This type of support is crucial for resolving issues promptly and maintaining the data center's uninterrupted operation.

The commissioning documents need to show more than just how the data center looks once it's built. They also need to include any changes made during both the building and testing phases. If something was done differently than what the original plans called for, that change must be explained clearly. The reason behind the change and its impact on system performance must also be included. This helps avoid confusion later on and ensures the information aligns with what was actually installed. Having this updated and complete record makes future planning easier and supports smarter decisions when it's time to make upgrades or improvements.

It's hard to overstate just how important good documentation really is. When everything is recorded properly, it helps avoid mistakes, makes troubleshooting easier, and keeps the data center running smoothly. Imagine what could happen if a key piece of equipment doesn't get its maintenance because someone missed a detail in the records. That one mistake could cause the equipment to break down unexpectedly.

That could result in major downtime and substantial financial losses. Or consider trying to fix or upgrade something, but without the correct drawings. That would lead to delays, higher costs, and frustration for everyone involved. These types of mistakes don't just cause short-term problems—they can significantly impact the long-term performance of the entire facility and compromise reliability.

The handover process from the construction crew to the operations team needs careful planning. It's not just a quick meeting—it's a full transfer of knowledge and responsibility. The construction team should walk the operations team through every part of the data center, explaining how each system works. This should be more than just a talk—it should include hands-on demonstrations and real training sessions. That way, the operations team doesn't just hear about the systems—they learn how to use and maintain them the right way.

The team also needs to go through every document together. The operations crew should have a thorough understanding of where the documents are stored, how to access them, and how to utilize them when needed. This includes reviewing test reports, as-built drawings, manuals, maintenance schedules, and other relevant documents. The operations team should receive training on how the document system works, enabling them to find and read information promptly. A checklist should help track what has been reviewed and ensure that nothing gets missed. This step is crucial to ensuring a smooth transition.

At the end of the handover, both the construction and operations teams should sign off on the process. They should agree in writing that the commissioning is complete and that the data center is ready to start running. The sign-off should include the date, the names of all parties involved, and any outstanding issues still pending resolution. This report is more than just a formality—it protects both sides and becomes an important part of the project's official record. It's the final step in a lengthy and detailed process, setting the stage for a stable, long-lasting, and reliable operation moving forward. Creating a detailed operations manual is a crucial component of the handover process. This manual should serve as a comprehensive guide to operating and maintaining the data center, providing clear instructions on daily operations, troubleshooting common issues, and performing routine maintenance.

The manual should be easily accessible to the operations team, ideally in both physical and digital formats, ensuring it is readily available in various circumstances. The clarity and comprehensibility of the manual are of paramount importance, as this document will serve as the primary source of information for day-to-day operations. Regular updates and revisions are crucial to maintain the accuracy and relevance of the manual, ensuring it reflects the current state of the data center and any operational changes that may occur over time.

The operations team requires a comprehensive training program that enables them to become familiar with the data center's systems, daily procedures, and the new operations manual. This training should not just focus on theory. It also needs to include real hands-on practice and drills with simulated problems to help build their skills and confidence. Each training session should align with the team's current knowledge and experience, ensuring that every team member can effectively handle their role.

This includes clear instructions on what to do in emergencies and how to manage a crisis. Providing them with these tools enables them to react quickly and effectively when something unexpected occurs or a serious issue arises. The goal is to build a solid and reliable team that knows how to keep the data center running smoothly, regardless of the challenges that come their way.

The success of the handover depends heavily on how complete and well-organized the documentation is. If the documentation lacks detail or structure, the operations team may run into confusion, make mistakes, or even cause downtime. Those kinds of problems are expensive and avoidable. That's why a clear, consistent documentation system matters so much. It supports the team in making fast, accurate decisions, especially during critical times. This system also plays a big role in long-term operations, helping keep things running smoothly for years. A clean setup now avoids trouble later on.

Along with training and documentation, the team needs a well-written operations manual they can use every day. This manual should include step-by-step guides, maintenance routines, emergency actions, and basic troubleshooting tips. It should be easy to read and always up-to-date. Creating and maintaining this kind of manual takes time, but it's worth the effort. It gives the team the backup they need when things go wrong or when new staff join.

Putting time and care into all aspects—from training to documentation to a comprehensive manual—helps avoid costly errors, builds confidence, and leads to a smoother handover. The goal is to give the operations team everything they need to run the data center without having to guess or figure things out on their own. When everyone understands the systems and knows where to find the right information, the transition from construction to operation becomes much easier. This meticulous process sets the tone for how the data center will operate for years to come. With the right tools and training in place, the team can maintain stability, resolve issues quickly, and minimize downtime. That's what keeps the facility reliable, efficient, and ready for whatever challenges might come next.

Vendor Coordination and Collaboration

Coordinating and working closely with vendors plays a major role in successfully commissioning a data center. This phase often brings together multiple vendors, each handling specific aspects of the data center's infrastructure. To manage all these moving pieces well, you need to take a proactive and organized approach that begins well before the actual commissioning starts. It all begins with a clear communication plan that outlines each vendor's role, responsibilities, and the process for escalating issues if problems arise. This plan helps ensure that everyone is moving in the same direction and that any conflicts are resolved quickly before they escalate into bigger problems.

The communication plan should list contact people for every vendor so the operations team always knows exactly who to reach out to for any issue. It should also outline the preferred methods of contact—whether by phone, email, or a shared project management platform—and establish expectations regarding the expected response time, depending on the urgency of the issue. Having this structure in place means that the right people can address problems quickly, which reduces delays and interruptions. It's also helpful to hold regular meetings—weekly or every other week, depending on the project's complexity—to review what has been done, what remains pending, and any potential issues on the horizon. These meetings should always include key people from the operations team, the commissioning crew, and all the involved vendors. Taking good meeting notes and sharing them keeps everyone informed and accountable.

One of the most important parts of working with vendors is setting up clear service-level agreements, or SLAs. These agreements outline what is expected of each vendor, including the response time required for addressing issues, the timeframe for resolving problems, and the acceptable level of downtime. Everyone involved should sign these SLAs to avoid confusion later. These agreements protect the data center operator by providing a clear plan of action if a vendor fails to fulfill their end of the deal. Good SLAs cover everything from basic maintenance to emergency fixes. As the data center evolves, these agreements must be reviewed and updated to keep pace with any changes in operations or requirements.

On top of that, it's important to include real performance metrics in each SLA. These measurements help track the performance of each vendor and indicate whether they're meeting their commitments. You can't improve what you don't measure, so this kind of tracking gives you the insight needed to evaluate their performance and spot any areas for improvement. By keeping a close eye on this and checking in regularly, you make sure every vendor continues to meet expectations and helps keep the data center running smoothly without surprises.

Beyond formal service level agreements, staying ahead of potential problems through solid risk management makes a big difference. That means identifying potential issues that could arise during the commissioning process, such as late equipment deliveries, a lack of skilled personnel on vendor teams, or unexpected technical problems that slow everything down. For each risk you identify, create a detailed action plan that outlines exactly what steps to take to reduce the likelihood of the problem occurring or prevent it altogether. Picking the right vendors early on is also crucial. You need to do your homework thoroughly—check their technical capabilities, financial stability, and reputation. Always follow up on references and dig into how they performed on similar projects. Don't just go for the lowest price. Yes, cost matters, but experience, reliability, and past success are equally important. You need to know what each vendor can do, what role they'll play, and whether their work style and safety practices line up with your standards. Picking the right partners early saves you from major headaches later on.

One thing people often forget is ensuring that everyone knows exactly what they're responsible for. It's crucial to be clear about who owns which part of the commissioning process, especially when multiple vendors are involved in the same task. Create a simple table that shows the task, the owner, and any support provided, if applicable. Keep it updated whenever roles shift or new tasks arise. When everyone knows their job, there is less confusion, fewer delays, and greater accountability. Pair this with a regular update system—simple progress reports that keep both the operations team and the vendors in the loop. When everyone has the same information, trust is built, and things run more smoothly.

Getting all the vendor systems to work together as one is another major step. This isn't easy. These systems often rely on each other, and if you don't plan that out well, things can go sideways fast. You need a detailed plan for bringing it all together—laying out the steps, deadlines, and assigning responsibility for each part. As each piece is integrated, run regular tests to ensure it all works together. Don't wait until the end to check it. Catching and fixing problems early prevents you from encountering significant delays or unexpected failures. Be sure to document every test, including what you tested, the outcome, and any fixes you made.

These records are important, not just for the present, but also for long-term maintenance and troubleshooting in the future.

The importance of post-commissioning support cannot be overstated. Vendors should provide ongoing support and maintenance for the equipment and systems they have installed. This typically involves Service Level Agreements (SLAs) that outline the level of support, response times, and other performance metrics. It is crucial to ensure that these SLAs are comprehensive and address all potential scenarios, from routine maintenance to emergency repairs. Regular communication with vendors is essential to ensure they are meeting the obligations outlined in the SLAs. This might involve scheduled meetings, routine performance reports, and prompt communication in the event of any issues. This ongoing support minimizes disruptions and maximizes the operational efficiency of the data center.

Finally, the entire commissioning and handover process should be documented meticulously. This documentation serves as a historical record, aiding in future troubleshooting, maintenance, and upgrades. The documentation should encompass all aspects of the vendor collaboration, including the SLAs, communication logs, risk assessments, and the results of integration testing.

This provides a valuable resource for the operations team, ensuring a smooth transition into the operational phase. The comprehensive documentation forms the backbone of effective ongoing management, supporting efficient problem resolution and ensuring the long-term operational health of the data center. The detailed record-keeping is an investment in future efficiency and operational success.

Ensuring Regulatory Compliance

Following all regulations is absolutely vital throughout the entire life of a data center, but it becomes especially critical during the commissioning and handover stage. This phase marks the transition from construction to full operation, and any regulatory non-compliance can lead to significant setbacks—ranging from project delays to legal issues. These may include project delays, service interruptions, or even fines and legal issues. That's why it's so important to understand and manage compliance carefully during this phase, and this section looks closely at the most important parts of that process.

Compliance should not be an afterthought. Start reviewing applicable codes and regulations early—ideally during the design stage—so you can build the data center correctly from the outset. Waiting until construction is finished can lead to costly delays and fixes. Compliance typically involves building codes, fire safety laws, electrical standards, environmental regulations, and data privacy laws. Staying on top of these rules helps avoid major issues down the line. Building codes are foundational because they cover the structure of the building, how people can access and exit, and the safety of working inside. These codes ensure the staff's safety and maintain everything to the highest standard. If the center doesn't meet these codes, you may face fines or even have to shut down operations until the issue is resolved. Problems may include missing fire exits, inadequate emergency lighting, or ramps that don't meet accessibility standards. To stay compliant, architects, engineers, and inspectors must work closely together throughout the construction and commissioning phases. You also need to schedule regular inspections and maintain clear records of them. These steps confirm the site meets all necessary codes and help you avoid last-minute surprises.

Fire safety regulations for data centers are stringent, given the high value of the equipment and data stored within them. These rules specify the type of fire suppression system required, the location of emergency exits, and the materials to use in construction. Compliance means installing advanced fire detection and clean agent suppression systems that won't damage electronics. Regular system checks, maintenance, and staff training—especially fire safety drills—are essential to meet these regulations and prevent equipment damage, fines, or more severe consequences. Training is a must. For example, fire safety drills should involve everyone who works at the site so that they know the steps to take in the event of an emergency. Not following these rules can do more than just damage your equipment—it can void your insurance, result in large fines, or, in serious cases, even lead to criminal charges.

Electrical safety plays a crucial role in ensuring a data center operates smoothly and safely, and the rules that govern this are strict for a reason. These electrical codes establish standards for everything from wiring methods to grounding systems, and they require the use of protective devices, such as circuit breakers and Ground Fault Circuit Interrupters (GFCIs). Meeting these rules isn't optional—it's necessary to prevent electrical accidents and ensure the power supply remains reliable.

Overlooking them could cause serious trouble, such as fires, damage to expensive equipment, or even harm to people working on-site. So, during installation and testing, every detail matters. Electricians must follow national electrical codes, such as the NEC in the United States, and undergo routine inspections. It's also important to regularly test backup systems, such as Uninterruptible Power Supply (UPS) units and generators. These tests make sure they work correctly in emergencies. Keeping clear records of all installations—such as wiring layouts, test reports, and maintenance logs—provides proof of compliance and facilitates future troubleshooting or upgrades. Alongside electrical rules, laws that protect the environment are becoming increasingly important in the operation of data centers.

These laws encourage facilities to reduce their energy consumption, manage waste heat more effectively, and minimize their overall environmental impact. Since data centers consume a significant amount of electricity, they must find innovative ways to cool their systems and minimize waste. This might include using energy-efficient cooling systems or switching to renewable energy sources, such as solar or wind.

They also need to keep a close watch on how much energy they use. Additionally, any type of electronic waste or hazardous material must be disposed of properly. Many areas also have strict regulations regarding how wastewater is handled. For example, cooling towers must often meet specific environmental standards before being approved. Failing to follow these rules can result in substantial fines and legal consequences. That's why centers must utilize environmental monitoring tools, submit regular reports to the authorities, and actively manage waste materials from start to finish.

When it comes to protecting data, privacy laws are equally important—especially for centers that store and manage sensitive or confidential information. These rules say how data should be stored, used, and protected. They also ensure that no one gains unauthorized access to the data. To comply with these laws, data centers must utilize robust security tools, such as encryption and access controls. Laws such as the General Data Protection Regulation (GDPR) in Europe and the California Consumer Privacy Act (CCPA) in the U.S. are strict and carry significant penalties for non-compliance.

Failing to follow them could hurt not only your wallet but also your reputation. Therefore, regular security checks, employee training, and clearly defined plans for responding to potential issues are all essential. During the commissioning phase, the team must also verify that every component of the security system functions properly and integrates seamlessly with the other systems on site.

Aside from these specific regulations, broader safety rules also play a significant role. Keeping workers safe—from construction through full operation—is just as critical. This means providing workers with the right training, ensuring they use Personal Protective Equipment (PPE), and adhering to safe work practices every day. It also includes conducting regular safety walkthroughs, audits, and providing clear reporting of any accidents or near-miss incidents.

Following OSHA standards in the U.S. (or similar agencies in other countries) helps reduce the chances of someone getting hurt on the job. You'll need to hold regular training sessions, ensure staff wear the correct gear, maintain a safe workspace, and have clear plans in place for handling emergencies. Job Hazard Analyses (JHAs) can also help identify and mitigate risks before they escalate into serious issues. Taking the time to do all this right not only keeps people safe—it also keeps projects on track and operations running without unnecessary setbacks.

You need to document every step of the regulatory compliance process with care and accuracy. These records should clearly show all permits you've secured, every inspection you've gone through, and the test results you've collected. This type of documentation provides proof that you've followed the rules and becomes especially important if someone needs to audit your systems or investigate a compliance concern. Keep all this information well-organized and easily accessible to those who need it—whether they're internal stakeholders or external authorities. Using a dedicated software tool can make it much easier to manage everything efficiently.

The records must be detailed, clear, and complete enough to track your progress, identify any issues, and quickly locate the necessary documents during an audit or emergency. Keeping all of this up to date and reviewing it on a regular basis helps you stay ahead of issues and makes the handover to full operations a lot smoother. It also helps everyone stay on the same page and reduces the risk of missed steps or overlooked problems.

However, following regulations doesn't end once the commissioning phase is complete. You have to keep working at it. This means scheduling regular audits, conducting inspections, and updating your systems and practices to ensure they continue to meet current laws and codes. Rules can change, so it's essential to stay informed about these changes. Don't wait until you're forced to catch up—keep your team informed and ready to adjust before it becomes a problem. Make this an ongoing part of your operations.

This mindset helps build a strong culture that values safety, responsibility, and high standards. It's not just about avoiding fines or penalties, though it definitely helps with that. It's about protecting the long-term health and success of your data center.

Consider regulatory compliance as an investment in your future. It keeps your business running smoothly without unexpected setbacks, shields your reputation, and helps ensure that everything runs smoothly, year after year. When your team respects the rules and stays prepared, your data center becomes more reliable, safer, and better able to handle whatever comes next.

Developing a Comprehensive Training Program

Creating a solid training program plays a crucial role in ensuring the data center transition proceeds smoothly. This isn't something to just get out of the way—it sets the foundation for smooth operations, safety, and following all the rules. A well-structured training setup equips your team with the necessary skills and knowledge to effectively handle the real-world challenges that come with operating a data center.

That means fewer mistakes, shorter downtimes, and more reliable uptime. A well-prepared team helps keep everything running smoothly. This section guides you through creating a robust training plan, covering the key steps that matter during the handover and for the long-term success of your operations. No training program works without understanding what your team actually needs.

Begin by conducting a thorough examination of the skills your team may be lacking and how these gaps could impact your operations. That way, you can shape the training to close those gaps. To get this right, you need to understand how your data center systems work and what each person is responsible for. Talk to the people involved—whether they're part of the operations team, supervisors, or other support staff. Use those conversations to figure out what they know, what they need help with, and what strengths they already have that you can build on. This process should encompass a wide range of skills, including working with PDUs, troubleshooting network issues, managing HVAC systems, and adhering to proper procedures during emergencies. You'll also need to cover safety practices, how to use your monitoring tools, and ensure they understand any regulations that apply to the work they do day-to-day.

Once you know what kind of training your team needs, you can build a proper training plan that actually works. That means setting clear learning goals for each section, determining how to teach the material, devising methods to assess its effectiveness, and establishing deadlines to complete the task. Build the training program in stages, making it easy to customize based on the needs of each individual.

One person may need to learn how to use a certain system from scratch, while someone else may just need a refresher on processes they already use. Organize the modules so that each one builds on the last, helping your team learn step by step. If someone struggles in a particular area, you can provide them with extra help without slowing down the entire group. This kind of structure keeps everything clear, keeps learning on track, and helps your team feel more confident about handling real-life problems inside the data center.

Selecting the right training methods is necessary when trying to make the program work for everyone. People learn in different ways, so using a mix of approaches helps cover all the bases. You can include instructor-led classroom sessions, hands-on lab work, self-paced online modules, and practical on-the-job training.

Classroom sessions provide a focused environment where instructors explain the material, answer questions, and facilitate discussions. Lab sessions provide an opportunity for individuals to experiment with concepts safely before working with live systems, allowing them to build confidence through real or simulated scenarios. Online modules are also helpful—they're flexible, cost-effective, and allow trainees to learn at their own pace, revisiting the content as needed.

On-the-job training puts the knowledge into action, with experienced staff guiding trainees through real tasks. Combining all these elements into one program often yields the best results. You can also utilize vendor training videos and existing internal training programs to complement the learning experience and maintain consistency with the tools already in place. Strong training materials make the difference between a program that sticks and one that doesn't. Good materials should be clear, easy to follow, and useful for individuals with varying levels of experience.

Lay them out in a way that makes sense, with goals for each section, real-life examples, interactive exercises, and built-in checks to test understanding along the way. Diagrams, flowcharts, and videos make tough topics easier to grasp and keep the material from feeling too dry. Using case studies helps too.

For example, demonstrating how poor maintenance led to a cooling system failure can effectively illustrate why preventive maintenance is crucial. It's also important to tailor the training to your specific site—include the systems and equipment your team actually uses. Keep the materials current by reviewing them regularly and updating them as equipment changes, procedures evolve, or new regulations are introduced.

You also need to test how well the training worked. That means using a variety of assessments to make sure people not only understand the material but also know how to apply it. Mix written tests, hands-on tasks, simulations, and evaluations based on real-world performance.

For instance, you could conduct a practical test where someone needs to locate and resolve the issue during a simulated power outage or restore communication with the BMS after it is disrupted. Set clear standards for passing each test and provide feedback promptly so that people know what to work on. At the same time, recognize the individuals who go above and beyond—it motivates the team and shows that their efforts matter.

Even after the training wraps up, continue to work on making it better. Ask trainees and supervisors for feedback to find what worked and what didn't. Examine test scores and survey results to identify patterns and determine where the training can be improved. Keep checking the program regularly to ensure it still aligns with the systems you're using, meets current regulations, and accurately reflects the skills your team actually needs. Continue listening, learning, and refining the program so it evolves in tandem with your data center and keeps your team sharp, ready, and confident at every step.

Consider implementing a robust knowledge management system. This can be a dedicated platform or something simpler, such as a well-structured SharePoint site. The goal is to consolidate all training resources, Standard Operating Procedures (SOPs), system documentation, and compliance information in one easy-to-access location. When your team knows exactly where to find what they need, they can work more efficiently and with greater confidence.

It encourages a mindset of continuous learning and reduces delays caused by missing or outdated information. Ensure the system is well-organized, easy to use, and updated regularly. Everyone should be able to search for and find answers without frustration. Providing your team with the tools they need helps them feel supported and ready to handle challenges as they arise.

A solid training program isn't just about technical knowledge. You also need to focus on soft skills, such as teamwork, clear communication, critical thinking, and sound decision-making. These skills are particularly important in a data center, where teams often need to collaborate under pressure.

Add specific training modules for soft skills and make them interactive to enhance learning. Role-playing, group discussions, and problem-solving activities work well. For example, simulate a scenario where the cooling system fails, and the team needs to respond quickly. Practicing those moments helps people build confidence and learn how to support each other.

Strong communication skills enhance coordination between departments and facilitate the reporting of issues or collaboration with vendors. Quick problem-solving helps limit downtime, and confident decision-making leads to better outcomes when emergencies hit. These skills support every technical task and keep operations running smoothly.

Scheduling and delivering training the right way is just as important as what you teach. Plan training around daily operations to minimize disruptions. If possible, run sessions during slower hours or roll them out in phases. That way, you avoid pulling too many people away from their duties at once. Ensure you have sufficient instructors, equipment, and space to run the sessions smoothly.

Clearly explain the training goals, timeline, and expectations for everyone. People need to understand why the training is important and how it helps them perform their jobs more effectively. When you keep them informed and involved, they're more likely to engage fully. Track attendance and progress and document everything from schedules to training records. This provides a clear picture of how things are progressing and helps you demonstrate to auditors or regulators that you're doing things the right way.

To wrap it all up, building a great training program takes time, effort, and a long-term plan. It goes beyond teaching people how systems work—it's about creating a team that learns together, works well under pressure, and continually seeks ways to improve. When you invest in solid training, you prepare your team for whatever challenges they may face. You reduce the risk of mistakes, avoid costly downtime, and keep your data center running efficiently and in compliance. With the right approach, your team becomes your strongest asset. They'll carry the success of the project from the construction phase straight through to reliable, high-performing daily operations.

On-the-Job Training and Mentorship

On-the-job training (OJT) and mentorship play a crucial role in developing a robust training program for your data center team. While formal training—like classroom sessions or e-learning modules—lays down the basics and covers important theories and procedures, OJT takes it a step further. It gives team members a chance to apply what they've learned in a real working environment. It helps them understand the systems they'll use daily, along with the small details that often don't appear in textbooks or training slides. At the same time, mentorship provides team members with personal guidance from someone who has already been through the ropes. A mentor supports, teaches, and motivates. When you combine both OJT and mentorship, you speed up learning, help people feel more connected, and build a stronger, more prepared operations team.

To run effective OJT, start with solid planning. Identify what each team member needs to learn and determine where their skill gaps lie. This involves reviewing the responsibilities associated with each role and understanding what makes your data center setup unique. Don't just focus on technical knowledge. Think about how well someone communicates, works with others, or solves problems on the spot. For example, a new technician might need hands-on experience with PDUs and power monitoring tools. Meanwhile, a senior engineer could benefit from mentorship on making informed decisions during emergencies or system failures. Every role requires a different approach, so tailor it accordingly.

Once you know what to focus on, build a clear OJT plan. Lay out the skills you want each trainee to learn, set a timeline, and define how you'll measure progress. Be specific about what tasks the trainee will perform, what kind of guidance they'll get, and what success looks like. That structure keeps things organized and helps both the trainee and their mentor stay on track. Make time for regular check-ins so mentors can give honest feedback and help trainees stay motivated. Use those sessions to make small changes if the training needs to be adjusted. Keep it flexible but focused.

Choosing the right mentors makes all the difference. A good mentor knows the systems thoroughly, has strong experience, and can explain things clearly. Look for people who are dependable, patient, and approachable. A strong mentor not only shares their knowledge—they inspire others to learn, grow, and improve their work. Try to match mentors and trainees based on their working styles and communication preferences. For example, pairing a quiet but curious trainee with a patient, thoughtful mentor can lead to excellent results. Matching by learning style and personality helps the relationship succeed, which results in better training outcomes. Strong mentor-trainee pairs build confidence, pass down valuable knowledge, and help create a team that functions smoothly and supports each other every step of the way.

Rolling out the OJT program means staying hands-on with the trainee's progress from day one. Supervisors and mentors must remain closely involved, providing real-time feedback and ensuring the trainee feels supported at every step. The mentor should clearly explain each task, answer questions without hesitation, and offer help as needed. Everyone involved should work to build a space where asking questions feels encouraged, not uncomfortable.

Open conversations, shared experiences, and consistent guidance help create a learning environment where people feel safe making mistakes and growing from them. This kind of atmosphere drives both confidence and improvement. Trainees should feel empowered to step up, solve problems independently, and actively seek input to continually improve.

Frequent and structured performance check-ins are a big part of making OJT work. These reviews shouldn't just look at technical know-how. They also need to measure growth in communication, teamwork, and independent thinking. A trainee might be great at running diagnostics, but do they know how to talk through an issue with a team during crunch time? These kinds of soft skills matter just as much. Both the mentor and the trainee should participate in these evaluations. That way, you get a well-rounded picture of how things are going and what could use some fine-tuning. The feedback during these reviews should be clear, helpful, and focused on actions the trainee can take to improve their growth. Use what you learn in each review to tweak the training plan if needed. This makes sure the program stays in sync with how each person learns and improves.

An OJT program that hits the mark does more than just teach tools and systems; it also fosters a deeper understanding of the work. It helps shape a mindset of learning, curiosity, and teamwork that sticks. When done right, it helps teams communicate more effectively, solve problems more quickly, and keep things running more smoothly across the data center. To keep that momentum, review the training program regularly. Technology evolves fast, and the way you train people should reflect those changes. Stay up-to-date with the latest updates in your equipment, procedures, and compliance requirements. Continue to communicate with both mentors and trainees to identify what's working and what's not. Their real-world input is priceless when it comes to keeping the training sharp and meaningful. Here's a quick example to bring it all together. One large data center recently launched a complete OJT and mentorship program tailored to its incoming tech team. They developed a comprehensive training plan from the ground up, encompassing hands-on work with power distribution, HVAC controls, EPMS, and BAS troubleshooting, as well as everyday server maintenance tasks. They selected seasoned technicians to serve as mentors—people who not only knew the technology but also possessed the patience and communication skills to guide their newer teammates.

The results showed fast improvements, better teamwork, and smoother handovers during shift changes. The program didn't just fill knowledge gaps—it gave people the confidence and support they needed to thrive in their jobs.

The mentorship program consisted of regular one-on-one meetings between mentors and trainees, during which they discussed progress, addressed challenges, and shared valuable insights and experiences. Mentors guided trainees not just through technical tasks but also helped them with their overall professional growth. These meetings provided trainees with an opportunity to ask questions, clarify any confusion, and build the kind of confidence that only comes from support and honest feedback.

The program also gave trainees hands-on opportunities to work on real tasks while having their mentors by their side. This allowed them to apply what they were learning in a real-world setting, while still having someone there to guide them and answer questions. Regular performance reviews became a standard part of the process, where mentors provided clear and constructive feedback, highlighting areas that required further attention. These evaluations helped trainees grow steadily, while also enabling the program to improve over time.

The results turned out better than expected. The trainees quickly acquired the technical skills they needed and adapted to the data center's daily routine without any issues. The mentorship program brought the team closer together, improving their collaboration and communication during shifts. It wasn't just about knowledge transfer—it built trust between teammates. And the benefits didn't stop there. This setup made the entire operation smoother by reducing downtime and enabling teams to react more quickly during unexpected issues. Everyone became more confident, and the team became more efficient. The clear success of the program showed just how important it is to mix mentorship and OJT into a larger, well-planned training strategy that fits the real needs of the team.

To build on that success, the data center added a knowledge-sharing system using SharePoint to support both OJT and mentoring efforts. This online library provided everyone with access to essential resources, including equipment manuals, standard procedures, how-to guides, and useful reference materials. Trainees could access the information they needed at any time, giving them more control over their learning. This extra layer of support helped fill in the gaps when mentors weren't available and reinforced everything covered during training. It became a go-to spot for answers, speeding up learning and keeping everyone on the same page. By combining structured on-the-job training (OJT), personalized mentorship, and a user-friendly knowledge base, the data center created a powerful and flexible system for quickly and confidently getting new hires up to speed and into the rhythm of daily operations. The team felt better prepared, the work flowed more smoothly, and it set a strong example of what a smart, people-focused training program can really do.

This mentorship program worked well in large part because the company took the time to make the process clear and structured. They laid out exactly what they expected from both mentors and mentees, including their roles, responsibilities, and how success would be measured. Regular check-in meetings were part of the plan, providing mentors and mentees with the opportunity to share updates, discuss challenges, and set short-term goals. To help mentors succeed in their roles, the company also provided them with proper training and support. This training included techniques for better communication, handling disagreements, and providing feedback that is both helpful and honest without being discouraging. That extra step made a big difference in how well mentors were able to guide their teammates.

The program didn't stop at teaching job-related tasks. Mentors also helped their mentees develop their soft skills—skills such as working collaboratively with others, discussing problems, and remaining calm in stressful situations. These skills are just as important in a data center as knowing how to handle equipment. By helping mentees develop confidence in these areas, the program created workers who were well-rounded and ready to handle both the technical and team-based parts of the job. Over time, this approach paid off in many ways. People stayed with the company longer, teams worked together more effectively, and everyone collaborated more smoothly. It lifted the mood around the workplace and helped get things done faster and with fewer issues.

To ensure the program continued to function as intended, the company established a method to monitor its effectiveness. They gathered input from both mentors and mentees, asking for honest opinions on what was going well and what needed work. They also monitored key performance indicators, such as how quickly new staff members picked up skills and how satisfied they were with their jobs.

That feedback helped the company fine-tune the program so it didn't go stale or fall behind. They treated it like a living thing—always growing, always being adjusted to fit the latest needs of the team and the work environment. This ongoing review ensured the program remained useful, practical, and relevant to the teams' actual needs. A well-run mentorship program like this doesn't just train new hires. It builds a strong, skilled group of people who feel supported and ready to meet whatever challenges come their way. Such investment in people yields better results, more reliable operations, and long-term success for the entire organization.

Documentation and Knowledge Management

Training your staff is a significant part of preparing a data center for operations, but establishing a solid documentation and knowledge management system is equally important. Think of this system as the "brain" of the facility—it holds all the practical tips, experience, and operational know-how your team gathers over time. Without it, important knowledge can slip through the cracks, especially when employees leave, or the workload becomes overwhelming. A clean, well-maintained knowledge base helps keep the entire team on the same page. It reduces mistakes, saves time when things go wrong, and helps the team work more efficiently.

At the heart of this system is good documentation, as mentioned earlier, which is thorough, accurate, and easy to understand. That means going well beyond the usual as-built drawings or equipment manuals. You need a full set of records for every key system in the building. These should include detailed specifications, step-by-step instructions for operations, service intervals, and straightforward guidance on resolving common issues. Everyone on the team should be able to access this information easily, regardless of their job or experience level.

Here's what that documentation should include to really make a difference:

System Documentation:

For all major systems—power, cooling, fire suppression, security, building controls, and network gear—you should maintain schematics, wiring diagrams, part lists, load data, and performance readings. Store this information easily accessible location (e.g., a shared digital folder or database). The documentation should include steps for system failures, backup procedures, and recovery times. Keep these documents up to date, and use tools that track version history to ensure everyone works from the most recent information.

Standard Operating Procedures (SOPs):

SOPs are step-by-step guides for routine tasks, such as maintenance checks, alarm responses, system resets, and handling outages. SOPs should be written in plain language that is easy for anyone to follow. They should be easy to read, clear, and direct. Review them regularly to ensure they match your current setup, especially after equipment upgrades or changes in rules. Use these SOPs during team training to give staff a chance to practice and build confidence before they're put to the test during a real event.

Troubleshooting Guides:

These guides should help technicians quickly identify the issue and determine the necessary repair. A good guide uses flowcharts or step-by-step checklists to facilitate clear and concise instructions. It walks you through likely causes, the signs to look for, and the steps to take. It should also include tips to stop the same problems from happening again. Use real examples from past incidents to keep the guides grounded in what actually happens on the job. The more your team adds their lessons learned, the more helpful these guides become over time.

A good documentation setup, supported by regular training, doesn't just keep the lights on. It builds a reliable, flexible, and prepared team that can handle pressure and keep things running smoothly no matter what comes their way. With this kind of system in place, your data center stands on strong, lasting ground.

Best Practices:

Writing down best practices helps the team work smarter and keeps the operation running smoothly. These shared tips and lessons learned to demonstrate the most effective, safest, and cost-efficient ways to accomplish the task. This section should consolidate the team's collective knowledge over time—shortcuts that save effort, tips for avoiding common mistakes, and effective strategies for reducing risk or cost. Think of it as a guidebook written by the people who do the work every day. This not only helps your current team work better, but it also sets up future hires and projects for success.

When people leave or move to new roles, their experience doesn't walk out the door with them. Instead, it becomes part of a living resource others can use and build on.

Inventory Management:

Maintaining a clear and up-to-date inventory of all equipment in the data center is crucial. This includes serial numbers, current locations, last maintenance dates, and any previous repairs. Having all that info in one place means you'll spend less time searching for parts or figuring out what needs fixing. It also helps prevent surprises by connecting your inventory system with your maintenance calendar. When you link the two, you can stay ahead of issues instead of reacting to them at the last minute. The system should be easy to update whenever new equipment is added or upgrades occur.

Knowledge Management Tools:

The tools you choose matter just as much as the information you store in them. If your system is too complicated or slow, people won't use it. Stick with something simple, searchable, and accessible to everyone on the team. You don't need fancy software—tools like SharePoint, a good CMMS, or even a well-organized internal wiki can work really well. What matters most is that your system aligns with the way your team already works and integrates seamlessly with your other IT tools and systems.

There are a few key things that make a knowledge management system truly useful:

Accessibility:

Ensure that everyone who needs the information can access it promptly, whether they're working onsite or offsite, during regular hours or in the middle of the night. Cloud access or a secure company intranet can help with that. At the same time, limit sensitive content to those who are authorized to see it. Use smart access controls to keep things safe while still being practical.

Searchability:

If people can't find what they're looking for quickly, they'll give up and go without it. Ensure your system utilizes smart tags, effective indexing, and concise file names. A well-organized, easy-to-navigate system reduces frustration and boosts team confidence. Usability:

Don't overthink the layout or features. The best system is one that even the least tech-savvy person on the team can use without stress. Keep the interface clean, the folders clearly labeled, and the instructions short and direct. When the system is simple, people use it often—and that's what makes it valuable.

Regular Updates:

Your knowledge base isn't something you set up once and forget. It needs attention and care. As procedures change or equipment is updated, your documentation should also be updated accordingly. Establish a clear process for team members to submit updates and assign someone to review and post them. Make those updates part of your regular routines, not just something done when there's time. That way, your system stays fresh, useful, and trusted by everyone who depends on it.

Version Control:

Keep every document clear and up-to-date by using version control tools. These tools enable your team to track every single change, so you always know who made what change and when. If someone makes a mistake or something doesn't work out, you can easily roll back to a previous version. No one ends up working off outdated files or guessing which version is correct. Everyone stays on the same page, which keeps confusion to a minimum and productivity high. Version control also helps explain why changes were made, so decisions are easier to understand and follow over time. That transparency keeps your documentation clean and your team confident in what they're using.

Training:

Don't assume your team will just figure out how to use the system on their own. Make time to train everyone on how to find what they need, add updates, and utilize version control effectively. Walk them through simple tasks like searching and tagging files, uploading new documents, and knowing when and how to make edits. This training shouldn't feel like a chore—keep it practical and tied to the actual work they do. When your team knows how to use the tools, they'll actually use them, and that's what makes your knowledge system strong and dependable.

A well-documented and effective knowledge management system is one of the smartest investments a data center can make. It not only reduces errors and supports training, but it also builds a culture of continuous learning and improvement. A well-maintained knowledge base ensures consistency, boosts efficiency, and prepares your team to handle any challenge. Moreover, it's invaluable during inspections, audits, or when reporting to clients. It provides transparency, builds trust, and keeps operations running smoothly, saving time and reducing costs in the long run.

Simulations and Scenario-Based Training

Building on a strong base of documentation and knowledge sharing, the next key step is hands-on training, specifically through simulations and scenario-based learning. Reading manuals and SOPs is helpful, but it doesn't prepare staff for the pressure and unpredictability of working in a live data center. While reading manuals and SOPs is helpful, they don't prepare staff for the unpredictability and pressure of working in a live data center. Simulations bridge that gap by providing a safe, controlled environment where teams can practice real-life situations, make quick decisions, and gain confidence before handling actual incidents. This type of training reduces mistakes during emergencies and enhances daily operations' efficiency. Simulations and scenario-based training offer several significant benefits. First, they offer a risk-free way to practice dealing with tough situations. Teams can try out different strategies, learn from their mistakes, and refine their skills with guidance from instructors who are well-versed in the field. This type of practice is especially useful in major situations, such as equipment breakdowns, power failures, or security threats. The cost of running these training sessions, even those that simulate large system failures, is small when compared to the potential consequences of a real emergency.

Another major benefit is better teamwork and stronger communication. When something goes wrong in a data center, it's rarely a one-person job. People from different departments must work together quickly and efficiently. Simulations give them the chance to rehearse how they'll do that.

They can follow established communication plans, learn how to hand off tasks smoothly, and gain a better understanding of who handles what. This type of training fosters trust and enables teams to respond quickly and remain composed under pressure. For instance, a fire drill might involve staff from operations, security, and fire suppression teams all working together. Going through it ahead of time helps everyone understand their role when it really counts. These scenarios also sharpen problem-solving skills. Real incidents don't follow a script. They come with surprises and force people to think fast. That's exactly what simulations prepare staff for.

These exercises train individuals to focus on the most important details, prioritize tasks by urgency, and remain flexible as the situation evolves. For example, if a cooling system fails, it might not just be about fixing one unit. Teams may need to identify how it affects nearby systems, determine which equipment requires the most protection, and work with controls to maintain stability for clients. Effective simulations come in many forms. They range from simple tabletop exercises, where teams discuss hypothetical situations and plan responses, to advanced computer-based simulations that mimic the behavior of complex systems. The type of simulation used depends on the training goals, available resources, and the complexity of the systems being addressed.

For instance, a basic tabletop exercise might simulate a partial power outage in one part of the data center. The team could discuss how the outage might affect different systems, review contingency plans, and determine the best course of action to manage the situation and minimize downtime. These exercises not only reinforce current procedures but also highlight potential weaknesses or gaps in existing plans, leading to necessary improvements.

More advanced simulations may utilize specialized software designed to replicate the behavior of critical systems, such as power distribution, cooling, or electrical systems. These tools create a realistic training environment, letting staff practice responding to system failures and test different strategies without putting real systems at risk. For example, a simulation might model a cooling system failure, allowing trainees to practice emergency cooling procedures, identify potential bottlenecks, and determine the best approach to restart the cooling system in stages.

The success of simulations and scenario-based training depends on a few key factors. First, the scenarios need to feel realistic and directly connect to the kinds of risks and challenges the data center staff face every day. These exercises should stretch people beyond their comfort zones, forcing them to think critically and react under pressure. If the scenarios feel too easy or unrelated to actual problems, they won't do much to prepare staff for real emergencies. The training needs to match real-life situations, such as when multiple systems fail simultaneously. These simulations should match the experience levels of the people involved, gradually increasing the complexity as they get more comfortable.

Second, a detailed and honest debriefing after each training session makes a big difference. Instructors should offer clear and helpful feedback, highlighting what went well and explaining areas for improvement. These debriefings should feel like a conversation, not a lecture. Everyone should have the opportunity to discuss what they did, what they learned, and what they might do differently next time. Talking things through this way helps turn training into real improvement. Teams often discover better ways to handle situations, and those new ideas can become part of regular practice.

Third, using the right technology makes these simulations even more valuable. Tools like virtual reality (VR) help teams train in a space that feels almost real. VR lets staff walk through emergencies without risking anything, and they can repeat the experience as many times as needed to build muscle memory.

Simulation software can also recreate how actual systems behave under stress. This lets staff run through all kinds of complex problems without touching the live equipment. It's a smart way to test how well a team works under pressure and try out different responses without the fear of making things worse.

Training exercises also need regular updates. Technology is constantly evolving, and new threats emerge frequently. A good training plan needs to grow and change, too. Keeping the scenarios up to date means the team stays ready for whatever comes next. Ignoring that step renders even the best training ineffective over time.

Fresh scenarios that reflect current challenges help the team stay sharp and alert.

To sum it up, hands-on training with real-feeling scenarios isn't just helpful—it's essential. These sessions provide people with an opportunity to practice, make mistakes, and learn without the pressure of a live emergency. That kind of practice builds confidence, enhances teamwork, and helps everyone work more effectively together. When combined with effective documentation and knowledge sharing, this type of training transforms a good team into a great one.

Everyone wins from this effort. The operations team feels more confident and skilled, clients see more reliable service, and the company builds a stronger reputation. A well-prepared team reduces outages, handles surprises more efficiently, and keeps everything running smoothly. These simulated experiences help protect the business, improve the service, and support long-term success.

Ongoing Training and Skill Development

Beyond the early stages of training and simulation, maintaining a data center's optimal performance depends on an ongoing commitment to skill development and learning across the entire operations team. This goes far beyond just ticking boxes for compliance. It's a smart move that strengthens the long-term reliability and smooth operation of the facility. In a field where things change fast, sticking with old skills just doesn't work. Data center technology, tools, and strategies continue to evolve, so the individuals managing them must also continue to grow. That's why having a solid, ongoing training program matters so much.

There are several key areas that this type of learning should encompass. First, the team needs to stay current with changes in technology. The world of data centers continually sees the introduction of new hardware, improved software, and tools that promise smarter ways to manage systems. Teams must understand how to work with these changes.

That means reading industry news, attending trade shows and learning events, and participating in vendor-led training sessions when new systems or tools are introduced. These learning opportunities help the team stay sharp and learn how to apply new methods that improve daily operations. For example, when an advanced power monitoring system is installed, staff require training on how it operates, what it tracks, and how to troubleshoot it when issues arise. The same goes for tools like updated cooling systems or software-defined networking gear. Each of these changes brings new things to learn, and proper training helps everyone keep up and make full use of the technology.

But learning shouldn't stop with technology. The way people work—the actual procedures and workflows—also changes over time. Standard Operating Procedures (SOPs) should be reviewed regularly, and the team needs to learn about the new updates and how to apply them effectively. This might mean group workshops, quick online courses, or hands-on coaching from someone already familiar with the updates. This kind of training is just as important as learning about tech. Teams must also adhere to safety rules, industry codes, and security regulations. Data centers must adhere to strict rules, and it's everyone's responsibility to be familiar with them.

That's why it's essential to conduct regular refresher training. Safety drills, updates to emergency response plans, and walkthroughs of physical and digital security steps should all be part of the schedule. These aren't just routine—they help keep people safe and reduce risk. When the rules change, the training must also change.

The materials should be updated fast, and the team should know right away what's expected of them. Cybersecurity, for instance, is constantly evolving. Teams require regular updates on the latest types of attacks, how to identify risks early, and what to do in the event of a breach. Clear, up-to-date guidance helps staff handle those situations confidently and protect critical systems and data.

The choice of training methods should be diversified to cater to People who learn in different ways, so training programs must match different learning styles and preferences. While classroom training still works well, incorporating online tools, hands-on simulations, and virtual reality (VR) enhances learning by making it more flexible and engaging. These options help staff stay sharp and make learning part of their routine, not just a one-time thing.

Online learning allows people to progress at their own pace and log in whenever they have time, which works well for those with tight schedules or shift work. Simulations offer staff the opportunity to practice real-world scenarios without the pressure of actual consequences. They can make quick decisions, see what works, and learn from mistakes in a safe setting. VR training provides an even more hands-on experience, enabling staff to learn equipment layouts and procedures in a realistic yet safe environment.

When you combine these tools into a single program—utilizing classroom learning, online courses, and practical simulations—you create a stronger impact. This blended learning style keeps things fresh and helps people really absorb what they're learning. Regular tests and check-ins also help by showing what has been understood and what still needs further work.

For training to truly take hold, the workplace culture must support it. Leaders need to demonstrate that training is valued and effectively linked to promotions and career growth. When staff see that training helps them grow, they're more likely to take it seriously. That means supervisors must understand each team member's skill gaps and learning needs. During regular performance reviews, managers should discuss with staff what training they may need, what they would like to learn, and how to support their growth. Teams also need time and budget for this. Training takes time, and companies should make sure staff can attend without falling behind on their regular work.

Flexible schedules and the smart use of technology can help keep everything running smoothly while training is in progress. Additionally, providing people with a reason to join training makes a significant difference. Offering rewards, extra pay, or better job opportunities to staff who complete training helps maintain high motivation. Celebrating when someone completes a course or earns certification demonstrates that learning is valued. This kind of recognition boosts morale, and when people feel proud of their progress, they stay more engaged and build stronger, more cohesive teams.

Mentorship programs are another great way to share knowledge and grow skills. Pairing seasoned staff with newer hires builds teamwork and makes it easier to pass on best practices. People can learn more effectively when someone with experience guides them through the process. Encouraging experienced team members to run internal sessions or lead small workshops gives them the opportunity to share their knowledge while retaining it within the company. A shared knowledge base or internal wiki can also be helpful. When teams build and use these tools together, they create a system where knowledge doesn't walk out the door when someone leaves.

Giving regular feedback also plays a key role. Don't just look at what someone has done—talk about what they could improve and how they can grow. Feedback should include clear next steps, like signing up for a course, joining a new project, or learning from a mentor. These conversations help resolve small problems early and provide people with direction. Over time, they also shape a team that's always learning, growing, and improving how the data center runs day to day.

The success of ongoing training and skill-building depends on solid planning, a clear budget, and regular tracking of results. A good training plan should outline the learning goals, describe how the training will be delivered, define what success looks like, and specify how the resources will be utilized. This plan can't stay static. It requires regular updates to stay current with changes in tools, industry standards, and safety regulations.

Things move fast, and the training should move with them. To ensure the program is working, teams should track progress using clear and measurable metrics. That could include how confident staff feel, how much knowledge they retain, fewer operational mistakes, or faster response times during incidents. Watching these numbers helps managers fine-tune the program, improve its effectiveness, and maximize the value from the money and time invested in it.

To wrap it up, ongoing training and skill development aren't just something extra to think about—they're key to running a strong and dependable data center. When teams get steady learning opportunities, they stay sharp, adapt to changes quickly, and make better decisions. Investing in this kind of training builds a workforce that is familiar with the systems inside and out and can respond calmly under pressure. This creates a workplace where learning never stops, and people take pride in continually improving their skills.

A solid training program does more than keep things running—it strengthens the whole business behind the data center. The commitment to growing the team reveals a greater promise: to maintain uptime, support client needs, and remain prepared for whatever challenges lie ahead. Over time, the benefits add up.

You get fewer outages, stronger performance, and higher team morale. Staff feel supported, valued, and ready. Most importantly, the data center becomes more stable, secure, and reliable—something every client and business partner can count on. The payoff? It goes far beyond just numbers. It helps build trust, keep services running smoothly, and protect the long-term health of the entire operation.

Developing Standard Operating Procedures SOPs

Creating clear and reliable Standard Operating Procedures (SOPs) is essential for the efficient operation of a data center. These procedures form the foundation of daily activities, promoting consistency, reducing mistakes, and ensuring effective troubleshooting. Developing SOPs requires a detailed approach that involves collaboration between different teams and a strong focus on accuracy.

The first step is to identify all operational tasks, ranging from routine maintenance to handling emergencies. This involves reviewing any existing procedures and analyzing the data center's unique systems and requirements. Each task must be broken down into clear, step-by-step instructions with precise decision points. Avoid ambiguous language at all costs.

Write in simple, direct terms that staff with varying skill levels can understand. Minimize technical jargon and define any terms when necessary. Adding visuals, such as diagrams, flowcharts, and photos, helps clarify complex tasks, especially those involving intricate systems or equipment. For example, an SOP for a routine power system check should include specific steps, not vague instructions like "check power distribution." Instead, it should detail:

- Go to Gallery Room X.
- Locate Building X, Data Hall X, and Power Distribution Units (PDUs) X.
- Use Meter X at Location Y to verify the main power supply voltage and current.
- Compare PDU readings with thresholds in Appendix A.
- Inspect all cabling for loose connections or damage.
- If you find any issue, follow the escalation procedure in Section 3.5.

- Log all readings and observations in the daily logbook.

Including details such as equipment models, locations, and references to appendices enhances clarity and traceability. A diagram showing the layout of the power distribution system can further enhance understanding of the system.

Developing SOPs isn't a one-time task. Regular reviews and updates are essential to account for technological changes, evolving practices, and updated regulations. Set a consistent schedule for SOP reviews, such as annually or after major system upgrades. Regular audits can identify areas for improvement, ensuring procedures stay accurate and practical. This ongoing cycle of creation, review, and refinement ensures that operations run smoothly and efficiently.

In addition to routine tasks, SOPs must include detailed protocols for emergency responses. These should cover scenarios like power outages, fires, security incidents, or equipment failures. Emergency procedures need a clear chain of command, defined roles, and established communication channels. Protocols should be accessible to all staff, displayed prominently in the data center, and available digitally on an internal network.

Regular drills and simulations ensure staff are familiar with these protocols and know how to act quickly and effectively. After any real or simulated incident, review the response to identify what worked, what didn't, and what can be improved. Update the SOPs accordingly to incorporate lessons learned, ensuring they evolve with the data center's changing needs.

Effective SOPs include detailed preventative maintenance procedures. These outline schedules, methods, and assigned personnel for maintaining critical equipment in the data center. For example, an SOP for cooling system maintenance might include inspecting chiller components, cleaning condenser coils, checking refrigerant levels, and performing system tests. Preventive maintenance prevents unexpected failures, reduces downtime, and extends the lifespan of equipment. Establishing a maintenance schedule, supported by tools like CMMS (Computerized

Maintenance Management System) software, ensures consistent execution and provides alerts for upcoming tasks. This proactive approach ensures systems run smoothly and prevents unplanned disruptions.

Creating SOPs requires collaboration across teams. The operations team contributes knowledge of daily workflows and best practices. The controls team provides expertise on building automation and electrical power monitoring systems. The IT team provides insights into network infrastructure and IT equipment, while the engineering and maintenance teams share their expertise in mechanical and electrical systems.

Involving all relevant stakeholders ensures the SOPs are thorough, accurate, and tailored to the data center's specific needs. A centralized repository, such as a SharePoint site or dedicated document management system, is essential for easy access and version control. This system should maintain a historical record of changes, with clear version numbering and logs to ensure traceability and accountability.

Training staff on new SOPs is just as important as creating them. Training shouldn't involve just reading the documents. Hands-on exercises, simulations, and quizzes help staff fully understand and apply the procedures. Refresher training should be a regular part of the schedule to reinforce best practices and maintain staff competency. Training sessions should also emphasize reporting any issues or anomalies during procedure execution, ensuring continuous improvement in operations.

The success of SOPs depends on consistent monitoring and evaluation. Conduct regular audits to check compliance and identify areas for improvement. Actively seek feedback from staff to understand how effective and user-friendly the procedures are. This feedback loop helps pinpoint ambiguities or inefficiencies, leading to ongoing refinements and improvements. Data gathered through audits and feedback can track key metrics, such as Mean Time To Repair (MTTR) and Mean Time Between Failures (MTBF). These metrics offer valuable insights into the effectiveness of the SOPs and inform future improvements.

Continuous improvement ensures the SOPs remain adaptable to the changing needs of a modern data center. Well-crafted and regularly updated SOPs reduce downtime, improve efficiency, and ensure the reliable performance of critical infrastructure, making them a vital part of the data center's operations.

Documentation Management Systems

Creating and using Standard Operating Procedures (SOPs) is only part of the equation. The other critical component is a strong documentation management system. This system serves as the central hub for all operational knowledge, ensuring that information is always accessible, consistently updated, and easily searchable. Selecting the right system depends on factors such as your data center's size and complexity, budget, and existing IT infrastructure. A smaller facility might find a shared network drive with a clear folder structure adequate, while a larger enterprise data center will need an advanced Document Management System (DMS).

Several factors should guide your choice. Scalability is one of the most important. The system must be able to handle growth in the number of documents and users as the data center expands. Consider future needs—will the system effectively handle increased storage and user demands? Integration is another key factor. A good DMS should connect seamlessly with tools you already use, such as your CMMS, ticketing system, and network monitoring software. This integration creates smoother workflows, reduces data silos, and gives you a unified view of operations. Security cannot be compromised. The DMS must include strong security features to prevent unauthorized access. These might include encryption, access controls, and regular security checks.

Data center documentation often contains sensitive details about infrastructure, configurations, and procedures that could be misused if exposed. Security features, such as multi-factor authentication, role-based access control, and audit trails, ensure that your critical documents remain secure and are accessible only to authorized personnel.

User-friendliness is just as important as security. The system should be simple to use, even for people who aren't tech-savvy. A complicated interface that requires lots of training can discourage adoption and defeat the purpose of having a centralized repository.

Look for a system with a clean design and helpful features, such as powerful search tools, metadata tagging, and version control. User adoption also depends on effective training programs that demonstrate to staff how to utilize the system efficiently.

The cost of implementing and maintaining a DMS is a critical consideration. Evaluate the initial cost of software licenses, hardware needs, implementation services, and ongoing maintenance fees. Compare vendors and systems carefully, taking into account the total cost of ownership. Don't just focus on the upfront cost; factor in the ongoing expenses for maintenance, upgrades, and support. Making an informed decision that aligns with both your budget and long-term goals is essential.

Once you've chosen a DMS, implementation involves several key steps. Start with thorough planning and design. Clearly define your goals and requirements, map out the process, and identify key stakeholders. Create a project timeline and allocate the necessary resources. Proper data migration from existing systems is the next step. Carefully transfer documents from previous repositories, ensuring their accuracy and integrity. Develop and test a robust migration plan to prevent data loss or corruption during the transition.

After migration, user training becomes crucial. Conduct in-depth training sessions for all users, covering navigation, document handling, and security protocols. Hands-on training builds user confidence, ensuring they can use the system effectively. Remember, the success of the DMS depends on user adoption, so investing in proper training pays off in the long run. Ongoing support is equally important. Regular system maintenance, updates, and technical assistance are necessary to keep the DMS running smoothly.

A strong DMS provides more than just document storage—it offers significant operational benefits. Version control ensures everyone works with the latest approved documents, preventing confusion. Metadata tagging and advanced search capabilities make it easy to find relevant documents quickly, saving time and reducing errors caused by outdated information. Workflow automation streamlines processes such as approvals and document distribution, enhancing efficiency and reducing processing times. Integration with other tools, such as ticketing systems or monitoring software, provides a comprehensive view of operations, enhancing collaboration and informed decision-making.

For instance, a SharePoint-based DMS is a popular choice for many data centers. SharePoint's flexibility allows for customization to meet specific needs, such as creating custom workflows or integrating with other Microsoft tools. However, setting up and maintaining SharePoint often requires specialized expertise, and costs can rise with increased customization.

On the other hand, cloud-based solutions offer a wide range of features, from basic file storage to advanced DMS capabilities. These platforms typically include automatic updates, lower IT overhead, and improved accessibility. But relying on an external vendor introduces potential risks, such as vendor lock-in or service disruptions during maintenance.

On-premises solutions offer greater control and customization, but they require dedicated IT resources for updates and maintenance. Choosing between on-premises and cloud-based options depends on several factors, including budget, IT expertise, security requirements, and integration needs. A hybrid approach, combining on-premises and cloud solutions, may provide a balance between cost-efficiency and security. No matter the chosen system, rigorous change control procedures are essential. These procedures ensure all document changes are tracked, reviewed, and approved before being implemented. Maintaining a detailed change log, with records of revisions, dates, and approvers, adds accountability and transparency.

Implementing a detailed documentation management system is an ongoing process rather than a one-time task. Regular audits are essential to ensure the system meets organizational needs and that the information remains accurate, current, and accessible. Audits also help identify opportunities for improvement, such as refining workflows, strengthening security features, and improving user training programs.

Continuous improvement must be a central focus. Collect feedback from users regularly and incorporate it into system refinements. This keeps the system aligned with operational needs and ensures its continued effectiveness. The ultimate goal is to create a system that not only stores documents but actively supports efficient and reliable data center operations. A strong DMS minimizes downtime, mitigates risks, and enhances overall productivity.

A well-implemented documentation management system goes beyond being a simple repository. It is a strategic tool that directly impacts the efficiency and safety of the data center. Regular reviews, user training, and a commitment to ongoing improvement ensure the system remains valuable and effective.

Change Management Procedures

Efficient data center operations rely on more than well-organized systems. Without comprehensive change management procedures, even minor alterations to hardware, software, or processes can disrupt services. This section outlines how to manage changes effectively, focusing on documenting, approving, and implementing them while minimizing risks. The primary goal is to mitigate potential problems by anticipating risks, documenting every step, and applying controls to reduce the chance of negative outcomes. Compliance, whether internal or with industry regulations and customer agreements, must remain a priority.

A strong change management process starts with a clear and standardized request procedure. This procedure should define the steps for initiating, reviewing, approving, implementing, and completing a change. An essential component is a standardized form—either digital or physical—that includes key details, such as:

- A clear description of the proposed change.
- The business justification for the change.
- The potential impact on other systems or operations.
- The proposed timeline for implementation.
- The individuals responsible for implementation and testing.
- A detailed risk assessment.

The risk assessment plays a critical role in the process. It must carefully evaluate the potential consequences of the change. What are the risks? How likely are they to happen? What impact could they have? What steps can mitigate or eliminate these risks? This assessment should cover possible downtime, financial impacts, security concerns, and other relevant issues. A formal risk matrix can assign numerical values to both the likelihood and impact of risks, helping to prioritize them and allocate resources where they are most needed.

Once the change request is complete, it must go through a structured approval process. This review typically involves several levels of oversight, depending on the complexity and potential impact of the change. Minor changes may require approval from a shift supervisor or team leader. However, significant changes that affect critical infrastructure or multiple systems will need formal approval from a group of stakeholders. This group might include IT, controls, operations, and senior management. Clearly defining roles and responsibilities in the approval process ensures accountability and prevents delays.

The approval process must also be well-documented and accessible to all relevant staff to avoid confusion. For significant changes, formal meeting minutes should record the approvals granted, providing an audit trail for accountability and transparency.

Once a change is approved, a detailed implementation plan must be developed. This plan should specify every step, including timelines, resource allocation, and contingency measures. Testing the change before deployment is crucial to ensure it functions correctly and doesn't interfere with other systems. The implementation plan should also include a rollback procedure—a clear and detailed method for reverting to the previous state if the change fails or causes problems. This rollback plan must be as comprehensive as the implementation plan itself. Executing the rollback plan promptly is key to minimizing disruptions, which requires regular drills and simulations to train personnel for high-pressure situations.

A post-implementation review is equally important and often overlooked. After the change is implemented, a thorough review should evaluate its results. Did the change achieve its intended outcomes? Were there any unexpected issues? What lessons can be learned from the process? This feedback must be documented and used to improve the change management process, creating a cycle of ongoing enhancement. This iterative approach not only strengthens the system but also provides valuable data for future risk assessments, enabling better predictions and more proactive strategies.

The change management process should integrate seamlessly with the data center's broader management system. This ensures consistency, reduces data silos, and streamlines workflows. Tools like CMMS (Computerized Maintenance Management Systems) or similar platforms can track changes, manage approvals, and automate parts of the process. Automated notifications, for example, can inform stakeholders about change requests, approvals, and implementation status, minimizing delays and improving communication. Regular reports summarizing change management activities should be generated for management review, helping identify trends, potential issues, and areas for improvement.

Integration with the documentation management system is another essential aspect. Every change must be thoroughly documented and easily accessible to authorized personnel. Documentation should include the change request, approval process, implementation plan, testing results, and post-implementation review. Maintaining this information within the DMS ensures a complete audit trail, facilitates compliance, and aids troubleshooting. Version control is critical to ensure everyone works with the latest, most accurate information, thereby reducing errors caused by outdated documents.

Consider a hypothetical example: a data center upgrades its network switches. The process begins with a detailed change request outlining the need for the upgrade, the switches to be replaced, the vendor, and the timeline. A risk assessment identifies potential downtime, network disruptions, and compatibility concerns. After stakeholder approval, a step-by-step implementation plan is developed, including a testing phase and a rollback procedure. Once the upgrade is completed, a post-implementation review assesses downtime, overall impact, and lessons learned. The entire process is meticulously documented in the DMS for future reference and auditability.

Regular audits of the change management process identify areas for improvement and ensure compliance with policies and regulations. By embedding change management into the core of data center operations, organizations can reduce unplanned outages, prevent security breaches, and maintain compliance. The goal is not just to manage change but to shape it proactively, aligning it with organizational goals and driving operational excellence.

Maintenance Schedules and Preventative Measures

Proactive maintenance is essential for ensuring the smooth and continuous operation of a data center. Unlike reactive maintenance, which focuses on repairing problems after they happen, preventative maintenance seeks to address issues before they occur. This strategy reduces downtime, improves efficiency, and prolongs equipment lifespan.

A successful preventive maintenance program isn't just a checklist; it's a strategic plan that requires careful scheduling, precise execution, and ongoing improvement. At its core, it involves creating comprehensive maintenance schedules, following manufacturer guidelines, and implementing preventative measures tailored to the facility's needs.

An effective maintenance program starts with a detailed schedule that outlines the required tasks and frequency for each piece of equipment. This schedule should reflect the specific needs of the data center, considering factors like equipment age, usage patterns, and environmental conditions. For example, older equipment may need more frequent checks than newer systems, and equipment in high-usage areas might require more regular inspections and cleaning. Environmental factors like humidity and temperature also influence maintenance frequency. For instance, high humidity could necessitate more frequent cleaning of cooling coils to prevent condensation buildup and corrosion.

Developing such a thorough schedule often involves using a Computerized Maintenance Management System (CMMS). These systems simplify the management of maintenance activities, offering a centralized view of equipment status, scheduled tasks, and unresolved issues. A well-designed CMMS reduces manual effort, enhances accuracy, and streamlines reporting and analysis. By identifying trends and areas for improvement, the system aids in refining the maintenance program. Features like automated alerts and reminders notify maintenance staff about upcoming tasks, ensuring they are completed on time and helping to prevent equipment failure.

Manufacturer recommendations are another key component of a successful maintenance schedule. Every piece of equipment comes with specific maintenance instructions outlined in the manufacturer's documentation. Following these guidelines ensures equipment operates within its designed parameters and retains its warranty. Neglecting these recommendations can lead to premature failures, costly repairs, and potential downtime.

Maintenance teams must have easy access to a centralized repository containing all manufacturer documents.

This ensures they have the right information to complete tasks safely and accurately. Regularly updating this repository with the latest revisions or updates from manufacturers is essential to keep the maintenance schedule accurate and effective.

Beyond scheduled maintenance tasks, additional preventative measures play a key role in extending the reliability and lifespan of the data center's infrastructure. These measures target specific risks and address potential issues before they escalate. For example, regularly cleaning air filters in cooling units is essential to maintaining proper airflow and preventing overheating. Neglecting these filters can drastically reduce cooling efficiency, leading to equipment overheating and failure. Similarly, inspecting cable pathways and connections can identify hazards like loose connections or damaged cables, preventing outages before they occur.

Preventative efforts also extend to the power infrastructure. Regularly testing UPS systems, including conducting battery load tests, ensures these critical systems operate correctly and provide backup power during outages. Skipping these tests can lead to system failure during a power outage, resulting in significant downtime and possible data loss. Inspections of power distribution units (PDUs) are equally important, focusing on loose connections, overloaded circuits, or signs of overheating. Any detected power imbalances or deviations must be corrected immediately to prevent catastrophic failures.

Environmental monitoring is another critical aspect of preventative maintenance. This involves continuously monitoring key environmental factors, such as temperature, humidity, and airflow, within the data center. Deviations from ideal conditions can signal underlying problems that may lead to equipment failures or decreased efficiency. Automated monitoring systems, paired with alerts and notifications, provide early warnings, enabling teams to take proactive action before issues escalate.

These systems also help identify patterns and trends, supporting better decisions about maintenance schedules and preventative measures. Data from environmental monitoring should be carefully analyzed to uncover weaknesses and guide improvements in the data center's environmental control systems.

A preventative maintenance program is not a one-time setup; it requires ongoing evaluation and refinement. Regular reviews of the maintenance schedule, combined with feedback from staff, are vital for optimizing the program. Key Performance Indicators (KPIs) and Critical Performance Indicators (CPIs) such as mean time between failures (MTBF), mean time to repair (MTTR), and overall uptime provide valuable insights into the program's success. Analyzing these metrics over time reveals trends and highlights areas needing further attention, helping to reduce the risk of failures and downtime continuously.

Training maintenance personnel is another cornerstone of a strong preventative program. Training should cover every part of the maintenance schedule, including the use of tools, safety protocols, and troubleshooting techniques. As technology, best practices, and manufacturer recommendations evolve, the training program must adapt to keep staff prepared. Properly trained personnel are less likely to make errors, which lowers the risk of damage or downtime caused by mistakes.

Meticulous documentation is the final key to a successful program. Every maintenance task must be thoroughly documented, noting the date, time, type of maintenance, personnel involved, and any relevant observations. This creates a valuable audit trail for compliance and aids in troubleshooting future issues. Detailed records also provide data for planning and refining the maintenance program. A CMMS simplifies this process with built-in reporting tools, allowing teams to track and analyze data effectively.

Thorough documentation boosts accountability, supports efficient troubleshooting, and helps with future planning. A well-documented preventative maintenance program instills confidence among stakeholders and strengthens the overall resilience of data center operations, ensuring reliable and efficient performance.

Incident Management and Response Plans

Effective incident management isn't just about reacting to problems—it's a proactive strategy that ensures a data center runs smoothly even when faced with unexpected challenges. A strong incident management and response plan anticipates potential disruptions, sets up clear communication channels, and outlines procedures to minimize downtime and risks. This plan acts as the framework that guides the team during unforeseen events, enabling quick responses and efficient resolutions. Without a defined plan, crises can spiral into chaos, leading to significant operational and financial losses.

The foundation of a successful incident management plan is a thorough incident reporting system. This system should make it easy to report incidents of any severity. Specific personnel or teams must be assigned to handle incident reports, ensuring no issue goes unnoticed or undocumented. The reporting process should be simple and intuitive, allowing fast and accurate reporting, even in high-pressure situations.

A standardized form—whether digital or physical—can help facilitate this process. It should record essential details like the type of incident, when it happened, which systems or equipment were affected, and how operations were impacted. The system should also enable the attachment of supporting documentation, such as photos or videos. These materials are often invaluable for understanding the issue and identifying the best solution.

Once an incident is reported, the escalation process becomes critical. This process establishes a chain of command for escalating incidents based on their severity and impact. Clearly defining escalation levels and providing contact information for each stage ensures the right people are notified promptly and decisions are made quickly.

Escalation procedures should also include timeframes to ensure there are no delays in addressing the issue. For example, a minor problem like a flickering light might only involve the facilities team. At the same time, a major power outage would require immediate notification of senior management, tenants, and possibly external support teams.

Regular drills and simulations are essential to test the escalation process and ensure everyone understands their roles. These exercises help identify weak points in the plan and prepare the team for real-world scenarios. A well-rehearsed escalation process ensures efficient communication and swift action during emergencies.

The core of any incident response plan is its capacity to organize the team efficiently and direct actions in a clear, structured manner.

Procedures for resolving incidents should be suited to the types of problems most likely to happen in the data center. This can include steps for dealing with power outages, network issues, security problems, equipment breakdowns, and environmental challenges like high heat or humidity. Each set of instructions should indicate what to do, who is in charge of each step, and what tools or assistance are needed.

The language used must be simple and clear to avoid confusion or delays. Checklists and diagrams should accompany the instructions to ensure nothing is overlooked. These procedures need to be reviewed and updated regularly to reflect any changes in the system, tools, or methods of work.

Once the problem is handled, it's just as important to look back and figure out what went wrong. A full review should follow, checking the cause of the issue, what helped it happen, and how well the response worked. This involves compiling reports, system logs, and staff notes. The goal is to identify weak spots in systems or steps and explore ways to prevent similar issues. Writing down the review and noting the ideas for fixing things is key. This feedback helps improve the overall plan for dealing with incidents, making it stronger over time. These reviews aren't just about the past—they help prevent future risks.

Using the right tools in the incident response plan is crucial. One helpful tool is a centralized incident management system (IMS). This kind of system helps track and fix problems by sending alerts, monitoring progress, and generating reports. It can work with other tools like monitoring software and maintenance systems to give a full view of how the data center is running. When selecting an IMS, consider how well it integrates with the center, its ease of use, compatibility with other systems, and data security. The system must handle multiple issues simultaneously and protect sensitive data while allowing authorized access.

Clear communication is important at every stage of problem handling. The plan should lay out how and when updates are shared so everyone who needs to know is in the loop. This means giving regular updates to renters, vendors, and managers. Communication might be through email, phone, text, or special platforms. Sharing updates in a clear and timely manner helps lower stress and builds trust, which in turn reduces problems. It's also a good idea to run drills and practice responses to make sure the communication plan works well under pressure and that everyone knows what to do.

Putting together and carrying out a solid plan to deal with incidents takes teamwork across all groups in the data center. This includes IT, facility teams, security, and renters. Ongoing training and practice help everyone understand their roles and how to effectively address issues. The plan should be reviewed regularly and updated with new tools, systems, and best practices. Everyone who might use the plan should be able to access it easily, and it should be integrated into the main data center management process.

Lastly, the flexibility of the plan will determine how well it works. What solves problems now might not work down the line as the center changes. Reviews, follow-ups after issues, and team feedback are all necessary to know the areas for improvement in the plan. It should never stay the same for too long—it needs to be updated regularly to stay useful. A plan that doesn't change can become a problem, but one that's always improved can be a great strength.

Being ready and able to adjust helps the data center stay up and running, reducing problems and keeping things working smoothly. In the end, this kind of strong and steady response makes the data center more reliable and valuable.

Understanding Critical Power Infrastructure

Understanding the critical power setup inside a data center is key to keeping systems running and avoiding downtime. This setup is the foundation that supports all operations, making sure power stays on for sensitive equipment. If even one part of this setup fails, it can trigger a chain reaction of issues, resulting in significant disruptions and financial losses. That's why it's important for data center managers to know how each part works and to stay ahead with smart planning and regular checks.

A big part of this system is the Uninterruptible Power Supply (UPS). These units kick in when the main power goes out, keeping systems running just long enough to shut down safely or switch over to a backup generator. There are three main types of UPS units: online, offline (also called standby), and line-interactive.

Online UPS units constantly convert power, which keeps it clean and steady and provides an instant backup when needed. Offline types only turn on when power fails, which makes them cheaper but slower to react. Line-interactive models sit in the middle, offering faster switching and some power conditioning features. Picking the right UPS depends on how critical the equipment is, how much money you can spend, and how long the backup needs to last. Big data centers usually use a mix of UPS types to keep costs low but still get solid performance.

Choosing the right UPS means looking at a few important details. First, it has to handle the full power load for all essential systems and IT gear. This includes determining the highest level of power the gear will require, plus some extra room to accommodate future growth or sudden spikes. Next, the run time matters. This is how long the UPS can keep things powered if the main power fails. The right run time depends on how long it takes to switch to backup or shut things down the right way.

Finally, efficiency matters too. A more efficient UPS uses less energy, which reduces bills and helps keep costs in check. But even the best UPS won't help if it doesn't work when you need it. That's why it's so important to test batteries, stick to a maintenance schedule, and run regular load checks. These steps make sure the system can handle real-life emergencies and won't let you down when it counts most.

Generators play a vital role in keeping the power running when utility outages last longer than expected. While UPS systems are great for short-term backup, generators step in for the long haul. Most of them run on diesel or natural gas and are built to keep everything running until the main power is restored. To achieve this, the generators must be sufficiently large and powerful to support all the components in the data center, including servers, cooling systems, lighting, and any other critical systems. Choosing the right size involves looking at past outage records, thinking ahead about future growth, and adding a buffer to cover any surprises.

A solid fuel plan matters just as much as the equipment itself. The best generator won't help if it runs out of fuel. That's why there needs to be a fuel plan in place that includes scheduled deliveries, constant tracking of how much fuel is stored, and safety rules for handling fuel properly. Fuel storage tanks must be checked regularly, and delivery schedules need to align with the expected runtime of the generator during an outage. Safety gear, training, and clear steps for dealing with fuel also help reduce risks.

To keep generators ready, regular maintenance is a must. This includes checking the engine, inspecting fuel lines, and conducting load tests to ensure the system can handle the full demand when called upon. These checks should happen on a schedule and be logged to catch issues before they turn into bigger problems. Skipping these tasks can lead to serious breakdowns when you can least afford them. A generator that fails to start when needed is more than an inconvenience—it can bring the entire operation to a halt.

Power Distribution Units, or PDUs, deliver electricity from the UPS or generators are directly connected to the racks and devices that need them. There are a few types to choose from. Basic PDUs simply pass power along, whereas intelligent and metered PDUs provide significantly more control. Smart PDUs let you check power usage and environmental data remotely. Metered PDUs provide detailed reports that help teams manage power more effectively and plan for future growth. In larger data centers, these advanced PDUs are the smarter choice because they give you the detailed view needed to cut waste and make the most of available power.

To spot trouble early, power systems need strong monitoring. This includes sensors to track voltage, current, temperature, and humidity, plus software that pulls everything together in one place. When set up correctly, the system sends alerts, helps identify trends, and enables teams to fix issues before they cause downtime.

Preventative maintenance plays a major role in keeping the power system in a data center reliable. A strong maintenance plan includes scheduled checks, testing, and upkeep of all key power components. These tasks cover things like battery checks on UPS units, load testing on generators, and regular inspections of PDUs and other equipment that help distribute power across the facility. The timing of each task depends on the importance of the equipment and the manufacturer's recommendations. Keeping detailed records of all maintenance work and test results is just as important as the work itself. These logs help teams spot patterns, make better decisions, and get a clear picture of how healthy the power systems are.

Routine checks are only one part of the plan. Capacity planning is just as important. Power systems need to grow alongside the demands of the data center. This means looking ahead and figuring out how much power will be needed in the future. Managers need to look at what's coming—new servers, expanded workloads, or bigger cooling demands—and plan for enough power without going overboard. Overspending on systems that are too big is a waste, but falling short puts everything at risk. Good planning uses past data, future goals, and smart risk thinking to make the best decisions.

To wrap it up, managing power systems in a data center takes solid planning, careful setup, and steady attention. Every part of the system matters. You need to understand how it works, monitor it with the right tools, and follow a maintenance plan that identifies problems before they escalate.

When you stay ahead of issues, the risk of failure decreases, and the data center remains up and running. For facility managers, investing in reliable systems and keeping them in good shape isn't just a box to check—it's the foundation for strong performance and long-term success. A well-maintained power setup doesn't just protect equipment; it protects everything the data center supports.

Optimizing Cooling Systems for Efficiency

Keeping cooling systems efficient is one of the most important parts of running a data center. Good cooling affects how long equipment lasts, how well it performs, and the cost of running the place. With so many servers and network devices packed into tight spaces, the amount of heat they produce adds up fast. That heat has to be removed quickly and effectively. If cooling isn't handled well, it can cause hardware to fail, slow everything down, or even lead to serious outages. That's why data centers need strong cooling systems and smart maintenance routines to keep temperatures under control.

One of the most common cooling setups in data centers is the Computer Room Air Conditioner (CRAC) unit. These units are standalone machines that work by pushing chilled air into the room to keep the temperature down. They're fairly easy to install and maintain, which makes them a go-to option for smaller data centers or places with simple cooling needs. But CRAC units aren't always the most efficient option, especially in bigger data centers. They don't target heat directly, so the cool air gets spread out more than needed. This leads to the mixing of hot and cold air—what many call "short-circuiting"—and that hurts cooling performance and wastes energy. To optimize CRAC units, teams should regularly clean filters and coils, ensuring the airflow remains steady. It's also important to fine-tune temperature and humidity settings through close monitoring. Some setups use hot-aisle containment to keep hot and cold air separated, which makes CRAC systems more effective and helps save energy.

A more advanced way to handle cooling is with Computer Room Air Handlers (CRAH) units. These systems work differently from CRACs. Instead of using built-in chillers, CRAHs pull in outside air and cool it before spreading it through the room. In cooler climates, this method uses less energy and can save money. CRAH units can also move more air and are easier to adjust for larger or more complex environments. But using air from outside brings its own set of problems. It's important to make sure that the outside air is clean and safe for sensitive equipment.

This means using high-quality filters to remove dust, moisture, or anything else that could cause harm. CRAH systems also need more advanced controls to manage airflow, humidity, and temperature just right. If any of those get out of balance, cooling won't be as effective.

The placement of CRAH units and the movement of air within the room are vital considerations. Poor airflow planning can lead to uneven temperature distribution, with some areas overheating while others stay too cold. This not only causes energy waste but also risks equipment damage. Proper planning ensures that cool air reaches the areas that need it most, reducing hot spots and improving overall cooling efficiency. Choosing the right cooling method and maintaining it properly are key for the smooth operation of a data center and for managing power consumption effectively.

As data centers continue to grow and become more powerful, traditional cooling systems just can't keep up with the heat generated by all the high-performance servers. That's where Direct Liquid Cooling (DLC) and Immersion Cooling come in. These cutting-edge cooling methods are becoming more popular because they're way more efficient at handling heat. With DLC, cooling is directed straight to the components that generate the most heat, like CPUs and GPUs, making it far more effective than air cooling. Immersion Cooling takes it even further by fully submerging the servers in non-conductive liquids that absorb heat even better.

Cooling has always been one of the biggest challenges in data centers, especially as servers get more powerful and racks are packed tighter than ever. Critical Facilities Managers (CFMs) are increasingly recognizing the potential of these advanced cooling technologies. While careful planning is necessary before implementation, many CFMs view them as prudent investments for future operations. These systems contribute to lowering energy costs, reducing reliance on traditional cooling equipment, and accommodating more servers within the same physical footprint. Furthermore, CFMs acknowledge the advantages of meeting sustainability objectives and preparing for stringent energy regulations. With the rising demand for AI, cloud services, and edge computing, facility teams are actively exploring the integration of DLC and Immersion Cooling into their long-term strategies, starting from the design and commissioning stages.

Overall, these cooling technologies are transforming the construction and management of data centers. They offer superior performance, promote energy efficiency, and support future growth and expansion. As the industry progresses, a fundamental understanding of how these systems operate and their benefits are becoming increasingly important for all stakeholders, from engineers to operations teams. DLC and Immersion Cooling are evolving from innovative concepts to essential tools for maintaining efficient and reliable mission-critical environments.

No matter what kind of cooling system is used, it has to be planned out well to run efficiently. That planning starts early, during the design phase. The team must choose the right cooling method for their setup, keeping in mind the data center's size, the density of the equipment, and the local climate. One of the biggest parts of getting it right is air distribution. If air doesn't flow well, some areas can get too hot while others stay cool. That wastes energy and risks equipment damage. A smart layout, especially one using hot-aisle and cold-aisle containment, keeps cool and hot air apart. This helps direct heat back to return vents and sends cool air to where it's needed.

This containment strategy utilizes physical barriers to separate the hot and cold aisles. But for it to work well, it must be sealed properly and allow the air to move the way it's supposed to. If hot air slips into the cold aisle, the system will waste power trying to fix the imbalance. So, the setup must be airtight, and the airflow must be managed closely to stay efficient.

Keeping things running smoothly means watching them in real time. Sensors need to track temperature, humidity, and airflow across the whole facility. These sensors send data back to a central system that helps staff spot problems early. If something gets too hot or out of balance, the system should send alerts right away. Smart controls can also adjust fan speeds or increase cooling power automatically to keep things stable. Some setups even use predictive tools that look at past data to predict future trends and prepare ahead of time.

Maintenance also plays a huge role in keeping cooling systems efficient. This involves regularly cleaning filters, coils, and pumps, as well as checking for leaks, clogs, or signs of wear and tear. How often you do these checks depends on the environment and the type of equipment in use. Sticking to a regular maintenance schedule helps catch issues early, extend the life of the system, and keep things running without hiccups. Teams should keep good records of each checkup and fix, which helps spot patterns and improve how the system works over time.

Beyond the technical aspects, the human element plays a vital role in cooling efficiency. Properly training data center staff on operating and maintaining cooling systems is essential for effective management and error prevention. This training should include safe operating procedures, troubleshooting techniques, and best practices for maximizing energy efficiency.

Additionally, effective cooling systems depend on good operational practices, like proper cable management to prevent airflow blockages and ensuring enough space around equipment for optimal heat dissipation. Cleanliness is crucial; regularly cleaning the data center floor and preventing dust buildup greatly improves cooling efficiency.

In conclusion, achieving efficient cooling in a data center is a multifaceted challenge requiring careful planning, design, deployment, and ongoing management. The selection of appropriate technologies, effective monitoring and control, preventative maintenance, and well-trained personnel all play crucial roles.

By optimizing cooling systems, data center operators can minimize energy consumption, extend the lifespan of their equipment, and ensure the reliable and efficient operation of their critical infrastructure. Efficient management of cooling is not merely a cost-saving measure; it's a critical component of ensuring the overall reliability and performance of the data center. The proactive approach outlined here will help ensure consistent, efficient cooling, protecting the investment in the data center and the vital data it houses.

Monitoring and Control Systems

Running a data center smoothly depends not just on strong power and cooling systems but also on smart monitoring and control tools that help manage everything in real time. These systems provide staff with a comprehensive view of the infrastructure's performance, enabling them to identify and fix problems before they escalate. Even with top-notch power and cooling setups, things can still go wrong without proper monitoring. That's why these tools are so important—they help catch issues early, reduce waste, and keep everything running at peak efficiency.

At the core of this system are Environmental Monitoring Systems (EMS), Electrical Power Monitoring Systems (EPMS), and Building Automation Systems (BAS). Sometimes, both EMS and EPMS are combined into a larger platform called a Building Management System (BMS).

All of these tools work together to gather and track data using sensors spread across the facility. These sensors measure things like temperature, humidity, airflow, power usage, and even the level of dust or particles in the air. Larger or more complex data centers often require additional sensors, particularly in areas where heat accumulates rapidly. Some facilities break things down by zones and use multiple EMS platforms, all feeding into one main dashboard for easier control.

Choosing the right kind of sensors matters a lot. Sensors must give accurate readings and work reliably in tough conditions. Wireless sensors are easier to set up and move around, which is particularly helpful in older buildings where running cables can be a hassle. But with wireless, you have to think about signal strength and the chance of interference. Wired sensors are more stable and offer better data quality, although they're more difficult to install and require more planning.

No matter which type of sensor you choose, you need to test and calibrate them regularly. Without regular checks, the data can drift and provide an inaccurate picture of what's happening. Creating a maintenance schedule helps ensure every sensor functions properly. This schedule should include cleaning, testing, replacing damaged parts, and updating software if needed. All of this helps the team trust the data and take action based on actual conditions, not guesswork.

Once the data is gathered, the system sends it to a central monitoring platform. This software collects all the readings, analyzes trends, and shows the data through charts and dashboards. These dashboards help the team quickly see what's going on and where action might be needed. The platform should also come with alerts built in. If a reading goes past a safe limit— like if the temperature spikes or power use jumps—it will notify the staff right away. That way, they can step in quickly before a minor issue escalates into a major outage.

These platforms also help shift the team from reacting to problems to preventing them. By studying the data and spotting patterns early, the staff can fix problems before they show up. This kind of insight turns monitoring from just watching into planning, and that makes a big difference in how well the data center performs long term.

Advanced monitoring systems are becoming increasingly sophisticated, thanks to tools such as machine learning and predictive algorithms. These systems can analyze past performance data to identify patterns and use that information to predict when something might break. This gives facility teams a head start, letting them plan and fix problems before they happen. For example, a smart environmental system might notice that a CRAC unit is starting to show signs of failure based on how it's been performing. With that early warning, the team can schedule a repair or replacement at a time that doesn't interrupt daily operations. That kind of planning saves money, avoids major downtime, and keeps everything running more smoothly.

Smart systems don't just wait for limits to be crossed—they use past trends and even outside data to make predictions. For instance, by studying patterns in temperature and humidity levels and comparing them with local weather forecasts, the system can figure out how much cooling will be needed on a hot afternoon. It can then adjust cooling levels in advance instead of waiting for things to heat up. This keeps the environment stable and the equipment protected while also cutting down on energy use. With power demands constantly shifting, this kind of flexible, real-time response is becoming more important than ever.

Getting a full picture of what's going on in the data center means connecting all the monitoring tools into one system. That means tying together environmental monitoring with power tracking tools. When these systems share data, they help the team see how different parts of the facility affect one another. For example, they might discover that one group of servers causes a spike in cooling needs at certain times. That information helps make smarter decisions about how to plan for future upgrades, move equipment around, or fine-tune cooling systems to use less power.

Tying everything together through the Building Management System (BMS) brings even more benefits. The BMS controls things like HVAC, lighting, and even security. When it works with the EMS and power monitors, everything can respond to changes automatically. For example, if the system detects a sudden increase in temperature, the BMS can activate additional cooling or redirect airflow to bring it down before any issues arise. This kind of setup doesn't just react to changes—it helps prevent them.

A fully connected system allows the facility team to be more proactive, not just reactive. With all this data in one place, they can plan better, save energy, avoid waste, and keep everything more stable. It turns scattered systems into a single, smart network that runs the data center more efficiently—and with fewer surprises along the way.

The successful implementation of monitoring and control systems. Setting up a reliable monitoring and control system takes careful planning and smart design choices. Choosing the right sensors and software should match the unique needs of the data center, including its size, layout, and how fast it's expected to grow. The system should be designed with growth in mind, allowing for easy expansion or upgrades when needed. Good documentation is a must. It should clearly explain how the system works, how it's set up, and how to troubleshoot any issues that may come up. On top of that, training the staff is just as important. Everyone involved needs to know how to use the system properly so they can get the most out of its features. While the tech does a lot, people still play a key role in making it all work. Operators need solid training, not just to use the system, but to understand what the data means. They must know how to read the numbers, spot warning signs, and act quickly when alerts show up.

It's not just about knowing the tools—it's also about understanding how things like temperature or humidity can affect the servers and other gear in real time. Running regular drills and practice scenarios helps prepare the team to act fast and stay calm when something goes wrong. That way, they don't just react—they respond with a plan.

Good communication is part of that plan. The team needs clear steps for who handles what and when, especially during emergencies. Setting up a clear chain of command and making sure everyone knows who to contact makes all the difference. These communication paths should include clear directions for escalating issues to the right people without delays. Every second counts when things start going off track.

In summary, effective monitoring and control systems are essential for ensuring a data center runs smoothly. These systems enable teams to anticipate issues, make real-time adjustments, and optimize energy use.

Troubleshooting Power and Cooling Issues

Troubleshooting power and cooling issues in a data center takes a clear, step-by-step approach. These systems are complex and often tightly connected, so a small issue in one area can lead to bigger problems somewhere else. Trying to fix things only after something goes wrong isn't just risky—it can be costly and cause significant downtime. That's why a proactive method, supported by regular maintenance and strong monitoring systems, is the better way to go. It helps avoid surprises and keeps the operation steady.

The first step in solving any power or cooling issue is to write everything down. You need to track exactly when the problem started, which parts of the system were affected, and what signs you noticed—like sudden power drops, overheating, or warning alerts from equipment. It's also important to note anything unusual that happened before the issue, which might have triggered it.

Keeping detailed logs across all systems helps identify patterns and provides a reference point if the same problem arises again. This it. This type of recordkeeping also supports effective root cause analysis, enabling you to identify and address the underlying issue rather than just treating its symptoms.

To dig into power issues, you need high-resolution monitoring tools that give accurate data on how much power each device and rack is using. These tools display current, voltage, and load values in real-time. If something looks off from what's normal, that's a red flag. For example, if one device suddenly starts using more power than usual, it could mean the device is malfunctioning, there's a wiring issue, or someone has plugged in something that shouldn't be there. By studying this power data and comparing it with environmental readings, you can often spot the actual cause. Smart monitoring platforms even use machine learning to find strange patterns that might be hard for a human to notice.

For cooling problems, you follow a similar method but focus on environmental data—mainly temperature, airflow, and humidity. EMS tools track these readings across the data center. If one spot is getting hotter than the rest, or if the humidity jumps in a certain area, that usually means something is wrong with part of the cooling system. Many cooling setups are designed with redundancy, which makes it easier to identify if a specific unit or section has failed. That way, you can fix or replace just what's broken without shutting down everything. Once you know something's wrong, the next move is to narrow it down and figure out what caused it. This often takes a mix of tools and hands-on checks. Start with a visual look—see if any cables are loose, broken, or look burned. Then test parts one by one, like checking a power supply, a cooling fan, or a temperature sensor to make sure they're working right.

Finally, go back to your data. Look at the trends, check what changed, and see what led to the issue. Root cause analysis helps prevent the same problem from recurring, saving time, money, and stress in the long run. Take the example of a server rack running consistently hot. At first glance, you might try improving airflow or cranking up the cooling. But a deeper look could show that the fans inside the servers aren't working properly, or the whole cooling setup was poorly designed to begin with. Fixing the surface issue might help for a while, but unless you solve the real cause, the problem will just keep coming back. Digging into the details—starting with symptoms and working all the way to the root cause—is how you make lasting improvements that keep things running smoothly.

The fix will depend on what the actual problem is. Sometimes, it's an easy job, like swapping out a bad part or tightening something that came loose. At other times, it may mean replacing a broken cooling unit or reconfiguring the power system. No matter the fix, take the time to weigh the risks first. Think about what might go wrong during the repair, how long systems might be down, what services could get interrupted, and if there are any safety concerns. Skipping this step could exacerbate the situation.

Once the work is done, test everything carefully. Don't just assume it's fixed—watch the system for a while to make sure it's steady and working like it should. Keep a close eye on key readings to spot any new or lingering problems. Write everything down—the root cause, what was done to fix it, and how the fix was tested. These notes are more than paperwork. They help you and your team build a solid troubleshooting guide for next time, saving time and reducing guesswork when similar issues pop up again.

Implementing a preventive maintenance program is another smart move. Catching small problems early keeps them from turning into big, expensive outages later on. A good schedule should include regular checks, cleaning, and performance tests for components such as power systems and cooling units. Assign these jobs to trained staff and ensure they adhere to the timeline. This doesn't just save money—it helps your gear last longer and keeps the whole system more stable. Make the tasks clear, the timing specific, and the responsibilities well-defined.

Training your team is just as important. People need to know what they're doing when things go wrong. That means regular sessions on how the systems work, how to identify problems, how to resolve them, and how to stay safe while doing so. Don't just stick to classroom learning. Run drills and practice scenarios so your team can get hands-on experience before the real thing hits. The better prepared your staff is, the faster and safer they'll fix things when the pressure is on.

When a power or cooling issue does happen, good communication makes all the difference. Everyone who needs to know—managers, IT teams, and even tenants—should get timely and accurate updates. Set up clear steps for who to contact, how to reach them, and when to escalate the problem if needed. Pick the best way to get the message out based on how serious the issue is. That might be an email, a quick message, or a phone call. When the team knows what's happening and what to do next, problems get solved faster and with less damage to operations.

Utilizing advanced monitoring systems is one of the most effective tools for resolving issues before they escalate. These platforms go far beyond simple alerts. They use predictive features to spot signs of trouble early. For example, they can flag hardware that's likely to fail soon, pick up on cooling systems running close to their limits, or catch signs of an incoming power surge.

Acting early on these warnings gives you time to fix the issue before it grows into a major disruption. The information these tools collect provides a clear view into how systems are performing and where potential issues may arise. That kind of data helps teams make better decisions, keep the equipment running smoothly, and avoid unexpected downtime.

At the same time, keeping detailed records during every step of troubleshooting is crucial. Every incident should be written down clearly—what the problem was, what steps were taken, and how it turned out. This kind of log doesn't just help with compliance and audits. It also serves as a valuable learning tool. Teams can look back, see what worked, and avoid making the same mistake twice. These records should be easily accessible to all team members for reference and review. As systems or procedures change, the documentation must also be updated. That way, everyone is always working with the latest information, and no one is left guessing during a crisis. This also makes handovers easier when new team members join or when shifts change.

Strong troubleshooting also depends on people—not just systems. Your team needs the skills and confidence to act quickly and effectively. Well-trained staff, paired with good data and smart planning, can turn potential disasters into minor blips. When people understand how to use the tools, recognize the warning signs, and follow well-documented processes, the whole data center runs better. This reduces stress on the team and improves the service for everyone who depends on it. But that takes regular effort—training has to be updated, new tools need to be learned, and habits have to stay sharp.

To wrap it up, solving power and cooling problems effectively isn't just about fixing things when they break; it's also about preventing issues before they arise. It's about spotting the early signs, learning from past mistakes, and making sure the whole team is prepared. Effective monitoring tools, accurate records, and a trained crew working together can make the difference between a minor issue and a major outage. Putting time and money into training, better tech, and strong documentation isn't just good practice—it's how you build a data center that people can count on every single day.

Capacity Planning and Future Growth

Effective capacity planning for power and cooling isn't a one-time task—it's a continuous process that requires regular updates, monitoring, and strategic adjustments. Poor planning often leads to power failures, equipment overheating, or disruptions that affect operations. A proactive strategy that addresses both short-term needs and long-term growth is essential. This section breaks down a clear approach to capacity planning that helps ensure your systems can handle what's coming next.

Start by auditing your existing setup—not just servers and network gear, but also backup systems, cooling units, generators, PDUs, and anything else supporting power and cooling. Document each component's full capacity, current usage, and available headroom. This forms the baseline for accurate projections. Simultaneously, monitor day-to-day energy consumption. Identify patterns in usage spikes and dips to better anticipate future demand. When combined with business growth forecasts, this analysis offers a clearer picture of what lies ahead. This is where DCIM tools really shine. A good DCIM platform doesn't just collect data—it gives you a window into how your systems are actually working. It tracks real-time numbers for energy use, cooling performance, and environmental conditions. It can show you where hot spots are forming or where airflow isn't doing its job. It points out areas where your systems are working too hard or not hard enough. You get charts, graphs, and dashboards that make it easier to spot issues and figure out what changes to make. These tools help you act early before something turns into a full-blown problem.

Picking the right DCIM tool takes some thought. key metrics that matter most to your team. Ensure it aligns with the way your facility operates and provides access to the key metrics that matter most to your team. But even the best tools won't help if nobody understands what the numbers mean. That's why you need team members who can dig into the data, recognize patterns, and connect what they're seeing to real-world decisions. Training in data interpretation and system behavior goes hand in hand with using the tool itself. The numbers are only useful when people know how to read them and what action to take. Strong planning relies on strong skills.

Planning for future power and cooling needs involves closely examining the type of IT equipment that will be added in the future. This type of forecasting is most effective when the critical facilities team collaborates closely with the IT department. Everyone needs to be on the same page. Forecasting models must consider factors such as expected server growth, upcoming hardware upgrades, and plans to introduce new technologies. Shifts like virtualization and cloud use also need to be part of the plan.

Virtualization, while efficient, increases power and cooling strain. Even with cloud adoption, hybrid and edge computing keep data centers relevant. Accurate forecasts depend on real-time data— not assumptions based on outdated models. This type of planning needs to stay flexible. It's not just a one-time estimate. Teams should run through a few different possibilities, including the best-case, worst-case, and most likely outcomes. This way, they'll understand the full range of what might happen and won't get caught off guard. Using sensitivity analysis also helps—this involves checking how changes in key numbers, such as server count or cooling efficiency, might impact the overall picture. This kind of in-depth look helps with budgeting and provides teams with a backup plan if demand grows faster than expected. For instance, if there's a sudden spike in the number of virtual machines, they'll already know what it might do to power and cooling needs and be ready for it.

Once forecasts are complete, build a concrete capacity plan. This plan should clearly lay out what needs to be done to meet future demands. That might include purchasing new equipment, upgrading older systems, or reconfiguring parts of the facility to improve efficiency. It should break things down into timelines and include the budget for each step. The plan also needs to weigh energy efficiency, not just cost. Choosing high-efficiency UPS systems, better cooling units, and smarter airflow designs will help reduce the center's energy bills and lower its environmental impact. These choices need to make financial sense over time, not just be cheap in the moment.

Putting the plan into action doesn't mean the work is done. It needs regular check-ins and updates to keep it aligned with what's really happening on the ground. As the business grows or technology needs shift, the plan must grow with them. Treat it like a live document that's always getting better.

This means adjusting when a new piece of equipment doesn't perform as expected or when the business heads in a new direction. Real-time data should be compared to the forecast often. That way, if something starts to drift off track, the team can act fast before it turns into something serious. This feedback loop is crucial for maintaining the data center's smooth operation and preventing unexpected issues in the future. Capacity planning involves more than just the technical side—it also has to take rules and regulations into account. Meeting legal and environmental standards is just as important as keeping the systems running.

This includes ensuring the plan adheres to all current building codes, energy efficiency laws, and environmental regulations from the outset. Leaving compliance out of the process could lead to delays, expensive changes, or even legal trouble down the road. That's why it's so important to include these rules as a core part of the planning process. Every upgrade, expansion, or change must fit within the limits of what the law allows. Teams should work with legal experts or compliance officers early in the planning stage to ensure that all steps remain on track and above board. It's much easier to build within the rules than to go back and fix problems after something's already been built or installed.

To wrap it up, solid capacity planning for power and cooling is key to a data center's long-term success. It's about more than just keeping up—it's about staying ahead. By taking the time to fully assess current systems, build accurate forecasts, and put smart plans in place, managers can prepare the data center for what's coming next without scrambling at the last minute. It's a process that takes regular checks, updates, and adjustments. It also depends on a clear understanding of how tech needs, business goals, and compliance rules all fit together. Sticking with this process over the long haul builds trust, reduces risk, and ensures smoother operations. A smart plan doesn't just support infrastructure—it empowers teams with the clarity and confidence to scale responsibly and sustainably. This approach protects uptime, keeps costs in check, and supports long-term stability, making the data center more prepared for whatever comes next.

Network Infrastructure Overview and Management

Bringing a data center from construction into full operation takes more than reliable power and cooling. A carefully planned and well-managed network is equally essential. This network acts as the center of everything—it moves data between servers, storage devices, and connects to the outside world. If the network operates smoothly, it maintains everything available, fast, and secure. That means business operations can continue uninterrupted, and sensitive data remains protected. To make this happen, the network must be carefully designed, and every part must be managed with meticulous attention to detail. This section covers the core components of the data center network and provides practical, real-world advice on how to manage and troubleshoot them effectively.

The network backbone begins with switches. Layer 2 switches handle internal data transfers between servers, storage arrays, and other devices. Selecting the right switches involves evaluating port speed (1GbE to 100GbE), port count, and advanced features like Power over Ethernet (PoE). As virtualization and cloud adoption increase, high-performance, non-blocking switch fabrics become vital for handling growing traffic without performance drops.

Equally important is cable management. Disorganized or damaged cabling can cause latency or service disruptions. Teams should implement a clear labelling system and conduct regular inspections for wear or loose connections. A documented cable plan—often stored in a DCIM platform—simplifies maintenance and accelerates issue resolution.

Then you have the routers, which work at Layer 3 and link the data center to other places, like the internet or a second data center. Routers move traffic based on IP addresses. Choosing the right router depends on the amount of traffic that needs to be moved, the type of security required, and whether the system will need to scale up in the future. For a large data center, strong routers with built-in security features are a must.

Setting up routing protocols like BGP or OSPF ensures that data takes the correct path to its destination. If these are set up wrong, traffic can get stuck, loop around endlessly, or crash the network. That's why using backup routes is a smart move. Protocols like VRRP or HSRP help here. They make sure there's a backup in place if a router stops working. To catch small problems before they become big ones, the team should keep an eye on the router's health. Watch CPU loads, memory use, and interface stats. Spotting unusual behavior early gives the team time to address it before things escalate. This type of proactive monitoring helps maintain network stability and prevents unexpected issues.

Firewalls safeguard the network by filtering incoming and outgoing traffic. Next-generation firewalls (NGFWs) provide advanced capabilities, including intrusion prevention, deep packet inspection, and application control. Proper configuration is crucial—missteps can create vulnerabilities.

Regular patching helps close emerging gaps. A single mistake in the setup can leave gaps in the system, providing attackers with a means of entry. Keeping firewalls updated with patches helps close those holes as they get discovered. When you pair them with Intrusion Detection and Prevention Systems (IDS/IPS), you add another strong layer of defense. These tools watch the traffic going through the network, looking for suspicious activity. If they spot something strange, they can jump into action to stop it. Where you place your firewalls makes a big difference. You want them guarding access to your most sensitive systems. It's also smart to keep an eye on firewall logs. These logs show what's been going on and help track down any odd behavior quickly. Checking them regularly means the security team can act before a problem becomes a real issue.

While switches, routers, and firewalls grab most of the attention, other network gear also plays a big part in how the data center runs. Load balancers, for instance, help distribute traffic across several servers so that no single server becomes overwhelmed. Network monitoring tools are also a must. They show what's going on in real time, helping the team spot and fix problems before users notice. For bigger setups, Network Management Systems (NMS) make life easier. They consolidate all network devices into a single view, allowing the team to monitor and manage everything from one location. Connecting NMS with other systems, such as DCIM, creates a comprehensive view of how everything works together. That big-picture view helps solve problems more quickly and keeps the data center running smoothly without unexpected issues.

Managing a network effectively requires ongoing effort. The team must continually monitor the situation, address issues before they escalate, and respond promptly when something goes wrong. Good monitoring tracks important metrics, such as the amount of bandwidth being used, the speed of traffic, and the frequency of errors. Keeping devices up to date with patches and firmware updates ensures they run safely and smoothly. When issues arise, using the right diagnostic tools helps the team identify the cause quickly and resolve it without delay. Having a clear plan for handling problems—who to call, what to do, how to communicate—makes a big difference during outages. Running regular tests of backup plans and disaster recovery steps makes sure the team knows what to do when something goes really wrong.

A strong security posture involves more than technology. Role-Based Access Control (RBAC) restricts system access based on user roles, and Multi-Factor Authentication (MFA) adds an extra layer of protection. Regular security audits, penetration testing, and policy reviews help identify and close gaps. User education is also critical—human error remains one of the biggest risks to network security.

Staying on top of updates, regularly checking policies, and monitoring new risks helps the team stay prepared for anything. A strong, steady approach to network security protects the entire data center and keeps operations running smoothly.

Planning for network capacity, just like with power and cooling, works best when done as an ongoing process. It's not something you do once and forget. You have to regularly look ahead and try to predict how much bandwidth the data center will need, especially as more servers come online and new applications demand faster speeds. These forecasts help prevent issues before they start. Regularly reviewing usage helps ensure the network can keep pace with growing traffic. If needed, upgrading to faster network technologies will maintain strong and reliable performance. It's also smart to line up these network reviews with power and cooling reviews, so everything grows together in the right way. A strong network plan should not only handle current needs but also be ready for what comes next.

This means thinking ahead about technologies like SDN (Software-Defined Networking) or NFV (Network Functions Virtualization), which offer greater control and flexibility. When the network design includes scalability and modular pieces, it becomes much easier to expand or replace parts without needing to rebuild everything from scratch.

To sum it up, the network is really the central nervous system of a data center. Without it, nothing moves, nothing connects, and nothing works right. A strong network that's planned out carefully, managed with attention, and built to be secure is key to keeping everything running without a hitch. That means planning, monitoring its performance, and making informed decisions about how to manage it every day. All of that helps the data center stay prepared for whatever the business needs next, from significant changes in technology to rising demand for speed and storage. The way the network gets managed, secured, and grown should always be at the center of the conversation. The right network infrastructure won't just meet today's needs. It will be ready for tomorrow's growth, tomorrow's technology, and tomorrow's challenges. When you plan and build it right, the network stays strong and better equipped to deliver consistent uptime, respond to change, and support the business well into the future.

Security Systems and Access Control

A strong network infrastructure, as discussed earlier, is only one aspect of maintaining a secure data center. Just as important—maybe even more so—is physical and logical security that protects the entire facility and everything inside it. A solid security strategy must employ multiple layers, from the methods used to access and exit the building to the tools that monitor for unusual behavior or attempted break-ins. You have to look at the different ways threats can show up and put the right defenses in place ahead of time. If something goes wrong, you should already have a plan in place to respond quickly and prevent it from spreading.

Physical security starts at the outer edge of the property. The first line of defense typically includes high fences topped with barbed wire or razor wire, accompanied by strong lighting to keep the area well-lit at all times. This makes it harder for someone to sneak in without being seen. It's also smart to reduce the number of access points so that entry and exit can be closely watched. Bollards can stop vehicles from ramming gates or walls. At each entry point, there should be systems, such as card readers or fingerprint scanners, so that only approved individuals can gain access.

These systems must connect to a central security platform that records every attempt to access the system. That way, if something unusual happens, the security team can review the logs and understand what happened. A good setup might also employ two or three layers of checks, such as a key card plus a PIN or a fingerprint scan, which makes it more difficult for someone to trick the system.

Once inside the building, tighter control is still necessary. Not everyone should be allowed to go everywhere. The level of access someone gets should match the work they do. For example, an IT technician might need to access the server racks, but a delivery worker shouldn't have that kind of access. These permission levels should be managed through dedicated access control software, often integrated with the building's management system. That way, you can easily track who's allowed where and make changes quickly when needed. It's essential to conduct regular checks on these permissions to ensure that people no longer have access to resources they no longer need. Every change to someone's access should be clearly recorded and reviewed for accuracy and security.

Maintaining access systems in good working order also requires effort. You have to regularly test card readers and biometric tools to make sure they continue to function properly. When someone leaves the company or changes roles, their key cards should be deactivated or replaced right away. The software that runs the whole access system also needs regular updates to keep it safe from new threats. Even the best tools won't help if they're outdated or malfunctioning.

In case of an emergency, people need to know exactly what to do. That means having written procedures in place for accessing restricted areas if systems go down. Those emergency steps should be tested regularly so that everyone is aware of what's expected. Contact details for the security team and clear instructions should be easily accessible, especially during a crisis. When people are aware of the plan and the tools are effective, problems can be addressed more efficiently and with less confusion, thereby protecting the data center and everything it supports.

Closed-Circuit Television (CCTV) systems play a key role in any strong data center security plan. A well-configured CCTV setup enables the security team to monitor all areas, both inside and outside the data center, in real-time. It also works as both a visible warning to potential intruders and a reliable tool for reviewing incidents after they happen. Placing cameras in the right spots is extremely important. You should avoid blind spots and make sure every entrance, exit, and high-value area has clear coverage. High-definition cameras help capture useful footage. You should also use either Digital Video Recorders (DVRs) or Network Video Recorders (NVRs) to store this footage. The recorded video can help during security reviews and serve as evidence if needed. Make sure to follow a clear retention policy that meets legal and industry rules. Test the CCTV system often. That means checking that cameras work correctly, recording devices have enough space, and footage is saved as expected. Skipping regular maintenance increases the chance of failure when you need it most.

To further strengthen physical security, Intrusion Detection Systems (IDS) utilize sensors to detect unauthorized movement or tampering. These can include motion sensors, door contacts, or glass break detectors placed in key locations. When the system detects unusual activity, it sends out alerts and activates alarms. Your IDS setup should work alongside your CCTV and access control systems to give you one complete picture of what's happening. Keeping a log of every alert helps you track issues and understand patterns over time. Test the sensors often and place them carefully to avoid too many false alarms, which can distract your team and slow down real responses. Make sure every team member knows the steps to take when the system sends an alert. Practice those procedures regularly to maintain sharp response times and prevent confusion during real incidents.

While physical security is critical, logical security protects your data center's internal systems from cyberattacks. Firewalls and

Intrusion Prevention Systems (IPS) help block threats before they reach your core systems. These should be set up carefully and kept up to date. Security should include multiple protective layers— this is called defense in depth. It's not just about technology. Your team requires regular training to identify phishing attempts, avoid risky behavior, and maintain safe online habits. Update all software and firmware often. This includes applying security patches to prevent known vulnerabilities from becoming entry points. Control who can access data and systems by following the rule of least privilege—only give employees the access they absolutely need for their work. Always use Multi-Factor Authentication (MFA), even for internal systems, to block unauthorized logins. Use security information and event management (SIEM) tools to keep all logs in one place. These tools let your security team detect threats early and react quickly.

All parts of your security setup must work together. This means integrating systems such as access control, CCTV, IDS, and network defenses into a single platform. This is typically accomplished through a centralized security management system (SMS), which collects data from all tools and provides a comprehensive view of what's happening. It also enables faster reactions and more informed decisions. Track security trends by reviewing reports and logs regularly. If something starts to change, this gives you time to act before the risk grows. You should also build a strong incident response plan and review it regularly. That plan should outline who is responsible for what during a breach, how to contain the problem, and how to restore everything to a secure state.

Clear, detailed documentation pulls everything together. Keep records on camera locations, sensor setups, access rules, and firewall settings. Update these documents anytime something changes. This helps everyone stay on the same page and allows quicker responses to problems. It also proves that you're following security rules and helps meet industry standards. Look over all your security policies often and update them as new threats emerge. Staying sharp and ready, with both planning and action, is the best way to protect not just your hardware but also the valuable data your customers and teams count on every day.

Data Center Security Best Practices

Building on the strong physical and logical security steps already mentioned, maintaining a secure data center requires a hands-on, layered approach that includes regular security checks, in-depth vulnerability scans, and a comprehensive program to train employees on security awareness. When these parts work well together, they really strengthen the center's protection and lower the chance of any break-ins or breaches.

Security checks aren't something you do just once and forget. They need to occur frequently and be part of an ongoing routine to ensure the center remains safe at all times. These checks should take place at least once a year—more if the center holds really sensitive data or faces bigger threats.

A skilled security expert with real experience in data center environments should take the lead. They should carefully review every system and process in place. That means ensuring that things like access control devices are functioning properly, reviewing camera footage to identify any unusual activity, and testing systems that detect break-ins. These checks should also cover software security, like firewalls, tools that stop hacks, and updated protection programs.

After the audit, the expert should compile a clear report that outlines the identified problems and the necessary steps to address them. The management team should review this report carefully and take immediate action. During the next round of checks, it's important to see if earlier suggestions actually worked. For example, if a past audit revealed that the outside lights weren't bright enough, a follow-up audit should determine if the new lighting or fencing has made things safer and more secure.

Running vulnerability scans is also a key part of the bigger security plan. Unlike regular audits, which examine what's already in place, these scans aim to identify weak spots before they become real problems. This typically involves using specialized tools that scan for known risks within systems and networks. These tools should scan everything—from servers to apps to devices on the network.

The goal is to find which weaknesses exist, how bad they are, and what damage they might cause. Once the scan is complete, a plan should be developed to address the issues identified, with the most urgent ones addressed first. It's essential to run these scans regularly, depending on the level of risk the company is willing to accept. For instance, a scan might show an old version of software running on a key server. Fixing that should be a top task to stop any future attack. These scans should also consider threats both from outside and inside, especially as cyber threats become increasingly complex and harder to detect.

People often overlook employee security awareness training, but it plays a crucial role in any robust security plan. Staff at the data center—from technicians to office workers—can often be the weakest link in the security setup. Phishing emails, clever trickery, and simple mistakes can seriously put the whole center at risk. A good training program should educate workers about various security risks and the proper methods for mitigating them.

This means learning how to identify scam emails, safeguard passwords, and follow proper procedures when handling data. Regular sessions help reinforce these points and remind people of the importance of staying alert and acting smart. Hands-on practice, such as simulated phishing attempts, reveals where employees remain vulnerable and provides valuable feedback for improving the training. It's also smart to shape the training around what each person actually does. Techs who work with servers may learn how to safely handle equipment, while admin staff may focus more on keeping private information secure and adhering to regulations. Every part of the training should be clearly tracked, with records showing who has completed it and who still needs to do so.

Additionally, implementing some extra best practices can further enhance the security of the data center. A solid plan for addressing security issues is crucial. This plan needs to explain exactly what to do if something happens—how to contain it, figure it out, and fix it. The team should run drills frequently to ensure the plan is effective and everyone knows their role. That could mean testing out fake attacks to see how the team responds and learning from the results. Managing outside vendors effectively also matters. Data centers often rely on third parties who need access to systems or the building itself. A good vendor program ensures that these individuals follow all security protocols. It's essential to conduct thorough background checks and closely monitor the activities of vendors while they're in the building or using any systems. Additionally, implementing multi-factor authentication at all entry points and systems helps prevent hackers, even if they manage to steal a password.

Keeping clean, up-to-date records also helps significantly in maintaining security. This includes documenting all policies, processes, and settings related to security. A well-organized system for documents helps ensure the team follows the rules and makes it easier to resolve problems quickly. These documents should be easily accessible and regularly reviewed to ensure they remain accurate, especially when circumstances change or new risks arise. Security training should also be updated with the latest threats and best practices. Building a strong security culture means more than just checking boxes. It's about getting everyone in the center to care about safety and take it seriously. Discussing risks and best practices should occur frequently through newsletters, team chats, or online lessons. It's just as important to create a space where people feel okay speaking up if they see something weird. Rewarding people who report possible problems can really help build a safer and more alert team.

To wrap things up, keeping a data center safe takes more than just strong locks and good software. You need a full-time, ongoing effort. Regular audits, routine scans, and real training for the team form the foundation. Implementing tested plans for emergencies, maintaining tight control of vendors, ensuring secure logins, maintaining a solid record-keeping system, and fostering a healthy security mindset all take it to the next level. Staying sharp and staying ready is the only way to keep both the hardware and the data safe. Constant checks, updates, and teamwork are the keys to keeping the center running smoothly while staying one step ahead of the threats that are always out there.

Network Monitoring and Alerting

Keeping a data center running smoothly depends on strong network monitoring and intelligent alert systems that detect issues before they become real problems. Without these tools, even small network hiccups can turn into big outages that cost money and damage trust. Modern data center networks are complex, with multiple layers like virtual machines, cloud tools, and all kinds of applications working together. Because of this, you need more than just a system that checks if things are up— you need a setup that truly understands how the network performs, where weak spots may appear, and how to catch and fix them before they affect any key services.

A good network monitoring system gives you a clear picture, in real time, of how well your network gear is working. That means everything—routers, switches, servers, storage—is constantly watched. The system gathers numerous data points, including the amount of bandwidth being used, the speed of data transmission, the number of packets being dropped, and the workload on the equipment (CPU, memory, and even temperature readings). What you watch depends on how important the device is to you. For example, you'd keep a closer eye on a server that runs a critical app than one that handles something small. All of this data should be easy to reach through a single dashboard, giving your team a full view of what's going on. This setup enables them to identify any unusual situations immediately and intervene promptly before issues arise. Showing key metrics with color-coded alerts, simple graphs, and clear charts helps everyone quickly spot patterns or slowdowns. It makes troubleshooting faster and way less confusing.

To really keep things under control, you need both hardware and software tools working together. You can start with basic network management systems, but more advanced tools go a step further. Some of these use artificial intelligence and machine learning to spot problems before they happen. These smarter tools learn from past data to identify patterns that indicate potential future issues. Let's say one tool sees a sudden spike in traffic on a server that normally doesn't get that much use. It can flag it early so the team can take action before it slows anything down. Picking the right tools for the job depends on the size and complexity of your setup, your budget, and the level of experience your team has in managing these tools. With some thought and planning, you can set up a system that covers what you need and avoids missing anything important. You just have to make sure the setup matches your data center's real needs and that the tools actually help the team stay ahead, not overwhelmed.

Maintaining a reliable network relies heavily on having a robust system for centralized logging and clear event analysis. Every network device should log important events, such as errors, security warnings, and any changes made to its configuration. A centralized log system consolidates these logs in one place, making it easy for the team to search, compare, and identify patterns. This makes it faster to find out what's causing a problem.

When done correctly, log analysis can help identify unusual behavior or trends that would otherwise go unnoticed, and this can reveal a great deal about the overall health of the network. For instance, if the logs indicate the same type of failure occurs every afternoon, it may suggest a traffic overload or timing issue. Using reliable log tools helps the team filter and analyze all the data efficiently without wasting time. They can identify specific issues, match up events, and determine what happened and why. That kind of insight saves time and prevents repeat problems.

Alongside logging, it's just as important to have a good system for alerts. The team needs to know right away if something serious goes wrong. A well-built alert setup informs the right people the moment a major event or error occurs. To make it work, alerts should trigger based on certain thresholds or patterns that you define ahead of time, so it focuses only on important stuff. You should avoid sending out too many alerts, or people might start ignoring them. That's why the system needs to follow a clear set of rules about which alerts matter most and how to handle less serious ones. A system like this can capture attention through various channels—email, text messages, or paging systems—depending on the urgency of the issue. When alerts come in too frequently or turn out to be false alarms, they only make things worse, so it's essential to fine-tune them for accuracy. With the right balance, alerts help people jump into action quickly without becoming overwhelmed by constant noise.

Real-life examples demonstrate the critical importance of this setup. Take a case where a server that runs a key app suddenly uses way too much CPU. If you've got a solid monitoring system, it will catch that spike and send out an alert fast. That message provides the team with all the key details—where the server is located, what the issue looks like, and its severity. With that info, the team can quickly restart the system, assign more resources, or get to the bottom of what's eating up the power.

Without alerts and tracking, nobody might notice until the whole server crashes and takes down services. Another example: a fiber cable gets damaged, knocking out access for multiple services. If your system detects the outage immediately, it will send an alert and share details, including the location and the number of services affected. The network team can arrive on-site, repair the cable, and restore services quickly. However, if nobody notices it until users start complaining, the fix will take longer, and the damage could be more severe. Creating an effective network monitoring and alerting plan requires time and careful planning. Start by picking out which parts of the network matter most. Some systems need closer watching than others. Set clear rules for what numbers or changes will trigger alerts.

Build out test runs that simulate real problems, like a server going offline or a drop in bandwidth. These drills help you see if the system works like it should. Run these tests regularly to fine-tune how everything runs, making sure you stay ahead of any weak spots. Good planning today prevents bigger headaches tomorrow.

Don't forget the importance of solid documentation. Write down how the system works—every part of it. That includes how you've set things up, what triggers alerts, and what to do when something goes wrong. This type of record helps bring new team members up to speed, facilitates troubleshooting, and enables you to catch any setup mistakes early. Keep these documents up to date by checking them often.

Every time you make a change in the system, reflect that change in the documentation. It's a living record, not a one-time job. A network that runs smoothly doesn't happen by accident. It takes effort, clear records, and a smart plan to handle changes over time. When you put the right tools, training, and structure in place, your network monitoring and alerting setup becomes the strong foundation that helps the data center stay stable, secure, and ready for whatever comes next.

Disaster Recovery and Business Continuity Planning

Disaster recovery and business continuity planning aren't just nice-to-have documents sitting on a shelf—they're must-haves for any data center, especially one that handles services and systems the business can't live without. If something big goes wrong—a flood, fire, cyberattack, or even just a power issue—it could cause major problems. We're talking about money lost, trust damaged, and operations brought to a halt. To avoid that nightmare, you need a clear, strong, and well-practiced Disaster Recovery (DR) and Business Continuity (BC) plan. This type of plan provides a step-by-step guide on what to do when things go wrong, helping your team recover quickly and get everything up and running again with minimal downtime.

To build a DR and BC plan that really works, you've got to start with a deep look at all the things that could go wrong. That means listing every potential threat that could impact the data center, assessing its likelihood, and determining the potential damage it could cause. You need to think about all sorts of risks—from earthquakes and hurricanes to equipment breaking down or someone accidentally deleting important data. Cyberattacks and even simple human mistakes also belong on that list. For every risk you find, the plan should include smart ways to reduce the chance it happens, and what to do if it does. You can't just make this list once and be done with it, either. Things change all the time—new tech rolls in, new threats pop up—so you should revisit and update this assessment at least once a year, and sooner if big changes happen in your setup or team.

A key part of your DR plan involves backups. These backups aren't all the same. How often you back things up should depend on how important that data is. If a system runs something critical, you may need to back it up every few minutes or even keep a live copy at another location. Other systems may not need that level of attention. A proven method called the "three-two-one rule" works well: keep three copies of your data, store them on two different types of storage, and keep one offsite. This way, even if your hardware crashes, a file gets deleted, or someone attacks the system, you've got something to fall back on. This level of backup coverage makes sure you're not left empty-handed if trouble hits.

However, saving data is only one side of the story—you also need a clear plan for retrieving it. Your DR plan should outline the steps to restore everything if an issue brings your systems down. It should explain the full process in detail, from who's in charge of what, to what tools are needed, and how long each part should take. It's not enough to just write it down and hope for the best. You need to run real drills to test everything. These drills should simulate various failure scenarios, including the worst-case ones, such as the entire site going offline. The goal is to see what works and where things fall apart. After the drill, take a good look at how your team did. What went right? What slowed things down? Apply what you learn to refine the plan and strengthen it. Don't forget to write down all the results and fixes, so you're better prepared next time. Keep practicing, keep improving, and make sure the plan evolves along with the rest of your data center.

Site redundancy plays a crucial role in ensuring the business continues to operate smoothly when issues arise. Setting up a second data center in a different area gives you a solid backup location for important systems and applications. If a disaster shuts down the main site, this second location can take over and keep things moving. Choosing the right location really matters. You want to avoid places that could face the same type of disaster as the main site, so don't put both in the same flood zone or earthquake region. The kind of redundancy you need depends on how much downtime your business can handle.

A hot site comes fully equipped and ready to run right away, but it's more expensive to maintain. A cold site, with just the basics, costs less but takes longer to set up if something happens. A warm site sits in between—partially ready, not as pricey as a hot site, and faster to launch than a cold one. What you choose ultimately comes down to your budget, risk tolerance, and how quickly you need to recover. It's also key to keep good records for the second site—what it can do, how to access it, and who to contact in case of emergency.

Implementing a solid disaster recovery plan involves bringing together various teams to collaborate effectively. It's not just the IT crew—it also involves the facilities team, security staff, and the business departments that rely on the data center. Everyone needs to know their role and what to do when the plan kicks in. You'll need clear lines of communication, written responsibilities, and effective coordination across all parts of the business to make sure things run smoothly.

Staying ready also means maintaining strong communication and keeping training up to date. The team needs to stay alert, and everyone involved must be aware of what to expect during an emergency. Talking with outside parties—like clients, partners, or regulators—is just as important. You need to give clear updates, manage expectations, and avoid confusion during a crisis to protect your reputation and keep trust intact.

Disaster recovery drills are about more than just testing the gear—they're about testing people and the whole process. These drills should simulate real problems, whether it's a small issue like a server going down or a full disaster that takes out your entire site. Run them often and make sure they include every step from the first alert all the way through restoring full service. The goal is not just to see if the plan works, but to find what's missing and fix it. After each drill, go through what worked and what didn't—both technical issues and communication problems. Write everything down, make updates, and continually refine the plan so it continually improves and remains ready for anything.

The plan should also spell out exactly how to protect your data when a disaster hits. You'll need clear steps to lock things down, keep intruders out, and make sure the data stays accurate and safe. This includes using encryption, implementing access rules, and adhering to secure cleanup procedures. If you face a cyberattack, the plan must explain how to contain the issue, recover systems, and conduct a review to identify what went wrong. Also, don't forget legal rules—make sure your plan keeps you in line with data protection laws before, during, and after a disaster hits.

But recovery is just one part. A good plan also keeps the business running while you make the necessary repairs. That might involve having staff work from home, relocating personnel to another site, or reassigning tasks. These options help keep important parts of the business going even when the main site is down. The plan should also outline strategies for maintaining contact with clients and others during a crisis. Regular updates and honest reports build trust and help avoid panic or confusion. At the end of the day, communication and teamwork drive the success of disaster recovery and business continuity. You can't just set it and forget it. Keep the conversation going—within the team, with vendors, and with customers.

Keep everyone informed and involved. This kind of open, honest, and steady flow of information fosters trust, prevents mistakes, and facilitates collaboration when things get tough. A well-planned, tested, and documented approach might not eliminate every problem, but it gives you the best shot at staying online, keeping your team safe, and serving your clients without skipping a beat—even during a crisis.

Establishing Clear Service Level Agreements (SLAs)

Creating clear and detailed Service Level Agreements (SLAs) is one of the most crucial steps in effectively managing tenant relationships within a data center. These agreements form the basis of understanding between the data center and its tenants, outlining expectations, defining responsibilities, and ensuring accountability on both sides. A solid SLA builds trust, prevents arguments, and keeps operations running smoothly for everyone involved. Ignoring this important part can lead to confusion, service failures, and, in worst cases, legal complications. Everything starts with a thorough understanding of what the tenant actually needs. That means having honest, open conversations— not just asking what kind of hardware they want, but learning how their business works, how much downtime they can tolerate, and what kind of impact a disruption would have. A company that does real-time financial trading needs a much stricter SLA than a small business that just runs an online store. The data center team must listen carefully and delve into the details to tailor the agreement to meet the specific requirements of each tenant. No two businesses are the same, so no two SLAs should be either. This process helps both sides get on the same page right from the start and prevents mismatched expectations later on.

Once tenant requirements are clear, those insights must be translated into precise, measurable commitments. Effective SLAs use the SMART framework: specific, measurable, achievable, relevant, and time-bound. Vague promises like "excellent uptime" are not enough. Instead, define commitments such as "99.99% uptime," along with what qualifies as downtime—whether it's a complete loss of service or extended performance degradation. Include Key Performance Indicators (KPIs) and Critical Performance Indicators (CPIs) such as network latency, bandwidth limits, and storage responsiveness. These metrics should be monitored using transparent, standardized tools, enabling both parties to verify compliance and maintain trust. Adopting industry-standard measurements also reduces ambiguity and strengthens the agreement's enforceability.

Support response times should also be clearly outlined in the SLA. Different types of problems need different levels of urgency. For example, a major system outage might require a response within 15 minutes, while a minor issue could wait a few hours. The agreement should also explain how problems get passed along to higher-level support if the first team can't fix them quickly. Everyone should know who's handling the issue and what happens next if it doesn't get resolved right away. It's also important to agree on how you'll communicate—whether it's through email, phone, or a support ticket system. And for each method, there should be a clear time frame for when the tenant can expect a reply.

Another key part of an SLA is what occurs if issues arise. The agreement should include a plan for service credits or compensation if the specified service levels are not met. This section should be straightforward and unambiguous, leaving no room for misinterpretation. For example, if uptime drops below the agreed threshold, the contract might specify that the tenant receives a certain portion of their monthly fee refunded for each hour the system is down. This approach demonstrates that the data center stands by its commitments and is ready to take responsibility if it fails to deliver. Clearly outlining how service credits are calculated helps prevent disputes. This transparency and fairness go a long way in building a strong, enduring relationship.

Besides the core points like uptime, performance, and support, a strong SLA should also include other details that matter to the tenant's daily operations. These might include access to the facility, procedures for requesting infrastructure changes, and details on how maintenance and upgrades will be handled. Planned maintenance should follow a schedule, and the SLA should explain how and when tenants will be notified ahead of time. The goal is to give enough notice to avoid disrupting their business. The agreement also needs to explain how a tenant can ask for changes to their systems—who approves them, how long it takes, and what costs could be involved. Covering all these smaller pieces helps everyone stay organized and keeps things running smoothly. It also fosters effective teamwork between the data center and the individuals it supports, resulting in a more positive experience for both parties.

Creating and finalizing the SLA should be a collaborative effort between the data center provider and the tenant. This back-and-forth process helps ensure the agreement truly reflects what both sides need and expect. It also lowers the chance of disagreements later. As the tenant's business changes or the technology they use evolves, the SLA should adapt accordingly. That's why it should be reviewed and updated often. These reviews can occur once a quarter or once a year, depending on the complexity of the services. Regular check-ins also give both sides a chance to talk about performance and fix small issues before they escalate into bigger ones.

Once both parties agree, the SLA becomes a legally binding document. It should clearly state what each side is responsible for and what they agree to provide. Everyone who needs access to the agreement should be able to find it easily, and the appropriate parties on both sides must sign it. Getting legal help when writing or reviewing the SLA is always a good idea. That way, you make sure the terms are fair and that the agreement protects both the tenant and the provider. This kind of careful approach helps build trust and makes the SLA more reliable, thereby reducing misunderstandings and potential conflicts.

The SLA also needs a clear plan for handling issues. If a problem arises, there should be specific steps to follow—starting with contacting support, escalating to higher-level staff if necessary, and possibly involving legal assistance or a neutral third party. Having a documented process like this ensures that minor problems are addressed promptly and fairly, rather than escalating into major issues. It also demonstrates that both sides value the relationship and understand that problems can be resolved without escalating more than needed.

Of course, a good SLA isn't just a bunch of promises written down. There has to be a way to track whether those promises are being kept. That's where proper monitoring and reporting tools come in. These tools should align with the performance numbers listed in the SLA, tracking them in real-time and generating reports that both parties can understand without requiring a technical background. Getting regular reports helps both the tenant and the provider stay ahead of issues. If something starts slipping, they can catch it early and fix it before it causes a problem. These reports also help identify long-term patterns, which can lead to improvements in performance over time. A clear system for checking and improving service adds real value to the agreement and ensures it remains useful.

Good communication is the heart of a strong SLA. It's what keeps both the data center provider and the tenant on the same page, helping everything run smoothly. Regular updates are key—they should include information about how the service is performing, upcoming maintenance, and any problems or unusual events that might affect the tenant. When both sides share information openly and clearly, it builds trust and helps avoid confusion. It also strengthens and enhances the working relationship over time, making it more productive and effective. These updates may occur during scheduled review meetings, through regular email communications, or via a shared online portal where both parties can view important details. Being open and prompt with communication helps catch small issues before they escalate into bigger problems, which keeps the relationship between both sides running smoothly.

To wrap things up, a clear and fair Service Level Agreement (SLA) is one of the most important aspects of keeping tenants happy and services running smoothly in a data center. Writing an SLA takes time and teamwork. The data center provider and the tenant both need to be involved in the process from the start. Planning carefully, talking openly, and working together ensures the final agreement fits the tenant's real business needs while also showing the provider's promise to deliver reliable service. But the work doesn't stop once the

SLA is signed. Keeping it alive through regular tracking, detailed reporting, and ongoing communication is just as important as writing it in the first place. These steps help ensure the agreement remains relevant and is followed. When these good habits are in place, both sides benefit. The tenant gets dependable service, and the provider earns trust and long-term loyalty. By following these simple yet effective best practices, data center operators can establish strong, enduring partnerships with their tenants—focused on reliability, shared goals, and consistent performance from both parties.

Communication Protocols and Reporting

Clear and open communication keeps a data center running smoothly, especially when you're working with several tenants who all have different needs and expectations. It's not just about reacting to emergencies—it's about having a strong, straightforward plan that keeps everyone informed, connected, and on the same page. This means setting up clear steps for sharing updates about system performance, maintenance plans, any scheduled interruptions, and any unexpected issues that may arise.

At its core is a solid communication plan. It should lay out which channels to use, how often to share updates, and who's in charge of sending what. This plan needs to show exactly how to handle different kinds of messages. That way, important alerts always go out to the right people at the right time. For example, if something goes wrong and systems crash, a quick phone call, text, or email should go out immediately. For less urgent matters, like upcoming maintenance, a simple email or a message through the tenant portal will do the job just fine.

A good tenant portal really makes things run more smoothly. It provides tenants with a single place to find all the important info they need—performance stats, maintenance schedules, ways to submit requests, and updates about issues. It can also store key documents like service agreements, user guides, and contact details. Features like automatic alerts for urgent issues, updates on service requests, and a simple ticket system improve communication, making it clearer and more effective for everyone involved.

You might also want to add a knowledge base inside the portal. It can answer common questions, guide tenants through basic tasks, and help reduce repeated back-and-forth questions. That saves time for everyone and keeps communication focused on the key issues that require attention.

The timing of updates should align with the needs of each tenant and the type of service they use. Some tenants might want updates every day, while others might only need them once a week or once a month.

A trading company working with fast-moving data will probably want updates in real time, maybe even through direct connections or custom data feeds. A hosting company, on the other hand, might be fine with less frequent updates as long as they're reliable. No matter how often you send them, the reports need to be clear, easy to read, and always ready to access. Good visuals, such as graphs, charts, or tables, can make the data easier to understand and more useful as well.

These reports should clearly include the key performance indicators (KPIs) and Critical Performance Indicators (CPIs) that matter most to each tenant, based on what is specified in their agreement. This could encompass factors such as uptime, network response time, available bandwidth, storage speeds, Power Usage Effectiveness (PUE), and temperature and humidity levels within the space. Including a history of these numbers, not just the latest ones, helps tenants see patterns, catch problems early, and avoid surprises. It also helps them make better choices about what they might need down the line. The reports should also call out any upcoming maintenance work. Providing tenants with ample notice allows them to plan and avoid last-minute issues. When communication is clear and early, tenants know they can count on the data center team. That kind of trust builds over time and really matters.

Besides regular reports, the team also needs to stay ahead of the game when something goes wrong. A robust system for handling problems must be in place to identify, understand, and resolve any service issues promptly and effectively. That system should work smoothly with the rest of the communication plan, so tenants get updates right away. They should hear what happened, what's being done to fix it, and when things should be back to normal.

Keep the messages simple and to the point. Avoid confusing tech terms, and focus on what the issue means for the tenant's setup. Even if things are moving slowly, a short update is better than saying nothing. Silence causes stress, and that's never good.

Open communication works best when it goes both ways. Setting up regular check-ins—either in person or through video calls— gives everyone a chance to bring up questions, talk through concerns, and plan. These meetings shouldn't just be for solving problems. They're also a chance to share helpful tips, look at future needs, and build a better working relationship.

Using various communication methods enhances overall effectiveness. This includes texts, emails, calls, the tenant portal, and social media (for non-urgent updates). Choose the method based on the situation and tenant preferences. For urgent messages, use multiple channels to ensure the message is received.

Choosing and using the right communication tools really matters. These tools must be dependable, easy to use, and capable of evolving to meet the changing needs of both the data center and its tenants. If the tools are clunky or break down often, they'll only slow things down and create frustration. Think about putting money into good monitoring and alert systems that send out alerts and reports automatically when something goes past a certain limit. These systems should work well with the tenant portal, ensuring everything runs smoothly and nothing gets missed.

Implementing a robust ticketing system can also significantly improve the handling of problems. With a shared system, everyone can see what's going on with each request—who's working on it, what's been done so far, and what still needs attention. That kind of setup builds trust because everyone knows things aren't getting lost in the shuffle. Additionally, reviewing past tickets helps identify patterns, which can reveal where recurring issues persist and where the process could be improved.

Keeping clear, detailed records is just as important. Every piece of communication—emails, alerts, meeting notes, support tickets—should be saved in one organized place. This not only helps solve future disagreements but also shows what's working well and what needs to change. Having everything documented means everyone stays on the same page and avoids guessing or making mistakes. A strong archive also gives a better picture of how things have changed and what's been learned over time.

It's also smart to check in regularly on how well your communication plan is working. Ask tenants and staff what's going well and where things feel off. Surveys, feedback forms, or even short chats can reveal a lot. Use that input to refine the plan and continually improve it. When the communication setup keeps up with what people actually need, the whole operation runs better, feels easier, and builds stronger connections with the tenants who rely on it every day.

Addressing Tenant Issues and Concerns

Managing tenants effectively requires more than just good communication—it also involves having a clear plan for addressing problems and concerns when they arise. Issues will occur, whether minor annoyances or larger disruptions, so it's crucial to handle them quickly and efficiently. Doing this correctly helps keep tenants satisfied and protects the data center's reputation for reliability and professionalism. This begins with establishing simple, easy-to-follow procedures for receiving, sorting, and resolving tenant problems from start to finish.

A strong ticketing system plays a big role here. Tenants should be able to log issues without hassle, see updates easily, and know where things stand at any time. The system should automatically route tickets to the right team based on the type of issue, so nothing falls through the cracks. Response times should match the seriousness of the issue. For example, a major outage that disrupts tenant operations should get immediate attention, while less urgent matters might follow a longer response window. Having clear service agreements that lay out timelines for responses and fixes keeps expectations realistic and fair. It also helps both sides stay accountable. Real-time dashboards and reports make it easier to track progress and spot problem areas before they grow. This kind of visibility gives everyone more control and improves the speed and quality of support.

When a ticket isn't resolved right away, the issue should follow a clear escalation path. Perhaps the front-line team can handle basic internet issues, but serious power outages require escalation to senior engineers or facilities leaders. Every step in the process should be well-documented, with specific individuals responsible for each stage. There should also be clear instructions on who gets notified, how they're contacted, and when that should happen. It's a good idea to review this process often to find what's working and fix what isn't.

Solving conflicts the right way is just as important. The goal should always be to listen carefully, stay calm, and work together to find fair solutions that help everyone. Blaming doesn't solve anything. Instead, teams should focus on understanding what caused the problem and how to fix it so it doesn't happen again. Sometimes, involving a neutral third party can be helpful when the issue is complex. Keep detailed records of these situations, including every step taken and what everyone agreed to. Those notes help clear up confusion later and guide improvements in how future issues are handled. Spotting issues before they turn into real problems is one of the best ways to cut down on the number and seriousness of tenant complaints. Staying ahead of trouble requires consistent system checks, effective maintenance planning, and a strong focus on preventing issues from arising in the first place.

For example, closely monitoring key data such as Power Usage Effectiveness (PUE) can reveal early signs of cooling system problems. That gives the team time to act before the issue starts affecting tenants. Routine checks of the physical setup—cables, hardware, airflow, and more—can also help catch small problems before they grow into big ones. These simple, steady actions show that the data center team takes quality seriously and builds real trust with the people who rely on them.

Staff training cannot be overstated in resolving issues effectively. It's not just about knowing how the equipment works. The team also needs strong communication skills, patience, and the ability to solve conflicts with care and confidence. Training sessions should teach skills such as how to truly listen, how to explain concepts clearly, how to remain calm in challenging situations, and how to resolve problems without exacerbating them. Holding regular refresher courses and offering opportunities for team members to grow helps ensure they stay sharp and ready to handle anything that comes their way. Using the right technology also makes a huge difference. Monitoring systems that track performance in real time help the team see what's going on and respond right away when something's off. These tools can send alerts the moment something unusual occurs, giving the team time to respond early. Analytics can also reveal long-term trends and patterns, making it easier to plan more effectively and avoid recurring problems. When used correctly, these systems help keep things running smoothly and prevent small issues from escalating into larger ones.

But technology alone isn't enough. Listening to tenants regularly is just as important. That means running surveys, holding feedback sessions, or just asking the right questions during regular check-ins. Honest feedback gives a clear picture of what's working and what's not. It also shows tenants that their opinions matter. Taking action on what they share—openly and quickly— goes a long way in showing that the data center team truly cares about doing better.

Creating a culture where solving problems early and working together are part of the daily routine makes everything stronger. Give the team the freedom and trust to handle issues as they come up, and encourage them to find creative ways to fix problems. Set up regular team huddles or knowledge-sharing sessions where everyone can discuss what's working and where there's room for improvement. When the whole team works together, learns from one another, and gets recognized for their efforts, the service only improves. In the end, solving tenant issues isn't just about fixing what's broken. It's about building solid, lasting relationships through action, communication, and trust. A strong, clear process that's backed by the right tools, a skilled team, and a real effort to listen and respond makes tenants feel valued. That trust keeps them around longer, builds a better reputation, and supports steady growth for the business. Checking in often, adjusting when needed, and prioritizing people keeps the strategy fresh and effective, regardless of how the data center landscape evolves.

Tenant Onboarding and Offboarding

Smooth transitions—whether welcoming a new tenant or managing a departure—are essential to successful data center operations. Both onboarding and offboarding require careful planning, structured processes, and consistent communication to prevent disruptions and ensure a seamless experience for tenants. Onboarding should begin long before a tenant ever sets foot in the building. It begins with a detailed conversation to determine exactly what the tenant needs. This includes power requirements, internet connectivity, cooling demands, and any specific infrastructure requests, such as raised floors or custom cabinet layouts. Obtaining these details early enables the data center team to prepare the space properly and avoid costly last-minute changes or delays.

Once requirements are defined, the next step is drafting a clear, customized Service Level Agreement (SLA). This document outlines mutual responsibilities and expectations—covering uptime targets, power and cooling expectations, security standards, and what happens if those promises aren't met. Make sure it's specific and leaves no room for confusion. Penalties for missed targets and a fair way to settle disputes should be part of it. After everyone signs off, a full project plan should be created to guide the setup of the tenant's space. That plan needs to include deadlines, key milestones, who's in charge of what, and backup plans for things that might go wrong.

Schedule regular check-in meetings to stay on track and quickly address any problems that arise.

Getting the space ready is one of the most important parts of onboarding. This may include pulling cables, installing PDUs, mounting racks, or modifying the layout. Every system— power, cooling, networking, and security—should be tested carefully to make sure everything works like it should. Keep detailed records of these tests to show the setup meets the SLA and follows best practices. Before move-in day, walk through the space with the tenant to fix any final concerns and make sure they're fully satisfied with what's been done.

When the space is ready, plan the actual move-in closely with the tenant. Set a clear schedule, assign move-in windows, and make sure everyone knows the steps involved. The data center team should be there during the move to help with anything needed. That kind of support helps prevent delays and reduces stress for everyone. Once the tenant is in, follow up right away. Have staff on hand to answer questions and fix any small hiccups that pop up. That personal attention after move-in shows that the data center team is committed to keeping things running smoothly from day one.

A smooth offboarding process matters just as much as a well-run onboarding. When handled properly, it protects both the data center's operations and the tenant's business from unnecessary disruption. To make this happen, the process needs to start early—usually several months before the tenant's planned moveout. The first step is simple: the tenant formally shares their departure date. Once that's done, the data center team can begin organizing the transition and assigning the right resources to manage each stage.

Next, the team should create a detailed decommissioning plan. This plan lays out every step required to disconnect and remove the tenant's equipment, clear the space, and get it ready for the next tenant. It should include a clear timeline, define who's responsible for each task, and address risks that might cause delays. A well-thought-out plan saves time, avoids confusion, and sets the right expectations for everyone involved.

The plan must cover all infrastructure details—from safely shutting down power and network connections to disconnecting cooling systems. Careful documentation is key here. Write down every task, from pulling cables to removing racks, and keep a checklist to make sure nothing gets missed. Holding regular check-in meetings with the tenant along the way helps keep things on track and allows both sides to raise and fix issues early. Before calling the job complete, do a final walkthrough with the tenant. This step confirms that all equipment has been removed, the space is back to its original condition (or whatever was agreed in the SLA), and nothing is left that could cause safety problems or delays for the next tenant.

Throughout this entire process, clear and consistent communication is non-negotiable. The tenant should receive regular updates on progress, and any issues that arise need to be addressed promptly. At the end, the data center should provide a full report that summarizes everything—what was removed, any final charges, and any open items. This protects both parties and keeps a solid record in case anything is questioned later. The team should also make sure any remaining data gets backed up and securely moved to the tenant's new environment, following all agreed security steps. Staff involved should understand data privacy rules and follow them without fail.

Handling onboarding and offboarding well takes more than just checklists—it also requires good tools. A solid Customer Relationship Management (CRM) system helps keep communication smooth and organized. It can track requests, deadlines, and who's responsible for what. When linked with work order tools and inventory systems, it keeps everything connected and easy to manage. Project management software also plays a big part here. It helps assign tasks, set priorities, and track progress, making it easier for everyone involved to stay in sync.

Using digital templates and checklists can really cut down on errors. They make the steps consistent every time and reduce the need for back-and-forth. Automated reminders and alerts keep everyone aware of deadlines and flag problems early, before they become bigger issues. These tools don't just make the work easier—they also collect data that can be used to spot problems, improve future processes, and raise the overall quality of service.

Reviewing this data regularly is key. The team should look at what worked, where delays happened, and what could be better. By doing this regularly, the data center remains flexible and continues to improve. This ongoing review helps the center stay current, meet new tenant needs, and avoid repeating the same mistakes. By examining trends, addressing bottlenecks, and remaining open to change, the team can improve their chances of making onboarding and offboarding processes smooth every time.

Both onboarding and offboarding require attention to detail, disciplined execution, and a proactive mindset. With clearly defined steps, reliable tools, and open communication, the data center can deliver consistent, high-quality tenant experiences. Beyond avoiding disruption, a structured transition process enhances reputation, boosts client satisfaction, and builds long-term operational strength.

Building and Maintaining Tenant Relationships

Strong relationships with tenants cannot be overemphasized in the long-term success of any data center. It's not just about offering rack space and connectivity—it's about building a true partnership based on trust, honest communication, and mutual respect. To build that kind of relationship, the data center team needs to stay ahead of tenant needs, respond to concerns quickly, and look for ways to add value at every step. That means reaching out before issues come up, not just reacting after the fact.

Communication works best when it's consistent and proactive. Instead of waiting for something to go wrong, the team should regularly share useful updates with tenants. A monthly newsletter can be a great way to do this. It could include news about system upgrades, upcoming maintenance, or any changes that might affect tenants.

To keep it engaging, add tips for improving security or saving energy, along with updates on industry events. Keep the layout clean, the language simple, and the content relevant. The goal is to keep tenants informed without overwhelming them.

Occasional extra updates might be needed if something major happens, but the regular rhythm of a monthly newsletter should be enough for most.

Face-to-face or virtual meetings, held every quarter, allow tenants to delve deeper. These meetings let everyone ask questions, give feedback, and hear what's coming up at the data center. They're also a good time to introduce new staff, celebrate wins, or explain changes to services. Open forums within these meetings give tenants a voice and help shape future improvements. They turn feedback into real action and show that the data center values input. To make sure that good ideas and promises don't get lost, the team should keep clear notes from each meeting, track decisions, and follow up on what was agreed upon.

A single, go-to contact for each tenant makes communication easier and faster. This person should handle everything from routine questions to urgent problems. With one clear point of contact, tenants know exactly who to call, and they don't get bounced around between departments. That builds confidence and speeds up problem-solving. To do the job well, the contact person needs support from the rest of the team, quick access to internal resources, and the power to make decisions when needed. Ongoing training and a solid understanding of the data center's communication plan help this person succeed and provide real value to tenants.

Good tenant relationships aren't just about talking—they're also about doing. Meeting the promises in the SLA is important, but great service goes further. It's about looking ahead and solving small problems before they turn into big ones. Keeping a close eye on environmental conditions, such as temperature and humidity, and ensuring backups are ready when needed, helps avoid downtime and demonstrates to tenants that their equipment is in good hands. Sticking to a regular maintenance schedule and notifying tenants in advance when work is planned helps build trust. No one likes surprises, especially when it comes to their infrastructure.

Going above and beyond creates a sense of partnership that tenants can count on. When the data center team consistently delivers high-quality service and communicates clearly, tenants feel heard and respected. That's what builds lasting connections. Even something as simple as being responsive and reliable during day-to-day interactions can make a big difference. Over time, these efforts lead to stronger tenant loyalty, fewer complaints, and a better reputation in the market.

By focusing on proactive communication, fast support, and thoughtful service, a data center can stand out. These small touches don't just improve relationships—they help secure the future of the business. Tenants who feel valued are more likely to stay, refer others, and grow alongside the data center. That kind of growth benefits everyone. Receiving regular feedback from tenants provides the data center team with valuable insights into how well things are working and where they can improve. Feedback can come in many forms—formal surveys, casual chats, or simple questionnaires. Well-designed surveys matter.

They should include open-ended questions along with rating scales to go beyond yes-or-no answers. This approach helps uncover both strengths and weaknesses, offering a full picture of the tenant experience. What's even more important is doing something with that feedback. Sharing the results and explaining the actions being taken builds trust, showing tenants that their input matters and leads to real change.

Building community among tenants also plays a significant role in enhancing overall satisfaction. Organizing events such as networking lunches, tech talks, or casual get-togethers provides tenants with an opportunity to connect, share ideas, and learn from one another. These shared experiences can lead to stronger tenant relationships, both with the data center and with other tenants. When tenants feel part of a supportive and collaborative community, they're more likely to stay long-term. Hosting workshops or inviting industry guest speakers also adds value and brings people together around shared interests. The staff can help by making sure everyone feels welcome and included at these events. Creating a space for communication doesn't stop with in-person events. An online community forum can provide tenants with a platform to ask questions, share their experiences, and offer advice.

When data center staff participate in these discussions, it shows they're listening and engaged. This type of forum can also serve as a useful resource, helping to build a stronger connection among tenants and between tenants and staff.

Resolving conflict is part of any business relationship, and the data center should be ready for it. Clear steps for handling disagreements help maintain fairness and consistency. A solid conflict resolution plan should guide how to respond when things go wrong—starting with open conversation and moving through a structured path if needed. Keeping all communication documented ensures transparency and protects everyone involved. When disputes are more complex, bringing in a neutral third party can help find solutions that work for both sides. The goal should always be a fair outcome and a stronger relationship at the end.

Helping staff handle these relationships well starts with strong training. Everyone should know how to talk with tenants respectfully, solve problems calmly, and stay professional when things get tense. Role-playing exercises give staff a safe way to practice tricky conversations before they happen for real. Offering regular training sessions and sharing new tips and best practices helps the entire team continue to improve its performance. Empowering staff to handle small issues on the spot—without needing to wait for a manager—also improves service and shows tenants that their concerns are taken seriously and resolved quickly.

Strong tenant relationships don't happen by accident—they grow from real effort, honest communication, and consistent service. Data center teams that listen closely, fix problems fast, and go the extra mile build trust that lasts. Creating a strong tenant community, following up on feedback, and continually seeking ways to improve helps keep tenants happy and loyal. That loyalty leads to longer contracts, fewer issues, and a better reputation overall.

Investing time and resources in tenant relationship management is a smart move that yields real returns. It keeps things running smoothly, improves day-to-day operations, and helps the data center stand out in a competitive space. At its heart, this work is about building partnerships, not just providing services. Trust, respect, and a shared focus on success turn everyday interactions into long-term, valuable relationships.

Understanding Data Center Governance

Running a data center smoothly takes more than strong infrastructure—it also takes a clear and well-enforced governance structure. This framework establishes the guidelines for approving, executing, and tracking work. It keeps things running efficiently, avoids unnecessary disruptions, and helps the team meet industry standards and legal requirements. Without a solid governance system in place, confusion and miscommunication can lead to costly mistakes and unexpected downtime. This section explains the most important parts of a strong governance setup and highlights the key roles that help manage work approvals.

Clarity around who's responsible for what is at the heart of good governance. When everyone understands their role, it's easier to avoid confusion and complete tasks effectively. A good structure starts with clear responsibilities and accountability at every step.

One of the primary roles is that of the Data Center Critical Facilities Manager (CFM). This person leads the whole process, makes sure procedures are followed, and signs off on any major changes. The CFM keeps everything aligned with the center's goals and risk prevention plans. They also help resolve conflicts and make sure tenants and other stakeholders stay informed and involved when needed.

Supporting the CFM, Assistant CFMs, and Chiefs work with a team of engineers and technicians who handle daily tasks. They review new work requests, put together detailed plans, and carry out the approved work. Their job includes ensuring that everything is done safely, efficiently, and on time. They also estimate how much time and what resources are needed for each job. Their expertise helps keep operations steady and prevents avoidable delays.

A strong governance system also includes a dedicated Change Management team. This team manages all types of changes using a structured process. They go through steps like submitting requests, analyzing potential impact, getting approvals, carrying out the work, and closing the change. They keep detailed records of every change made. These records help prevent problems, support audits, and make troubleshooting much easier if something goes wrong. A well-documented process also ensures that changes don't slip through unnoticed, which could affect operations.

When the work touches the network, servers, or software, the IT department gets involved. Their input helps ensure that the work aligns with the larger IT strategy. They also handle system configurations and updates, and they test everything to catch issues before they cause real damage. Their collaboration with facilities teams helps avoid downtime and keeps everything working as expected.

Tenant involvement is equally vital in a healthy governance system. Tenants should have straightforward ways to request work, receive updates, and voice concerns on anything that impacts their setup. Including tenants in this process fosters collaboration and helps prevent issues by identifying problems early and resolving them before they escalate. When the data center communicates transparently with tenants about what's happening and when, it builds trust and minimizes surprises. This openness enables everyone to plan better and reduces unnecessary stress.

By having the right people in place and ensuring each role is well-defined, the data center establishes a strong foundation for success. This type of structure ensures that things run on time, keeps tenants informed, and prevents costly downtime. It also ensures the team can respond quickly and clearly when something unexpected arises. Good governance doesn't just protect the data center—it also helps it run better every day. A clear process, backed by skilled people and good tools, sets the stage for long-term performance and tenant satisfaction.

The decision-making process for work approval plays a critical role in data center governance. Decisions must be transparent, fair, and based on clear criteria. Establishing thresholds for approval authority ensures efficiency and control. Minor changes can be approved at lower levels, while significant changes require higher-level approvals, potentially involving the CFM or the executive team. This tiered system ensures that decisions are handled at the appropriate level of responsibility without unnecessary delays.

Urgent requests, such as critical infrastructure repairs or security updates, must be addressed swiftly. A prioritization system helps rank requests based on their impact on data center operations and tenant business continuity. This system may employ a scoring method that takes into account potential downtime, the impact on business operations, and the resources required for resolution. Prioritizing requests effectively ensures that critical tasks receive immediate attention without compromising overall workflow efficiency.

A strong system for documenting the work approval process is essential. This should include all requests, approvals, work plans, and post-implementation reviews.

Comprehensive documentation serves multiple purposes: it creates an audit trail for accountability and compliance, tracks project progress, supports analysis of completed work, and helps identify potential risks. Utilizing a centralized document management system, such as SharePoint, enhances accessibility for all stakeholders and streamlines workflows, thereby promoting efficiency across teams.

An escalation process is equally important to resolve disputes or unforeseen challenges. This process should define clear steps for escalating issues to higher management levels, specifying roles and responsibilities. A transparent escalation procedure ensures accountability and provides clarity to all stakeholders, allowing conflicts to be resolved quickly and fairly. Establishing clear escalation guidelines helps prevent delays and ensures the smooth functioning of the data center.

Regular reviews and updates to the governance framework are crucial to maintaining its effectiveness and alignment with operational goals. These reviews should incorporate industry best practices, regulatory changes, and lessons from past experiences. Input from all key stakeholders—including the CFM, engineers, IT staff, and tenant representatives—is critical for refining the framework. Continuous improvements ensure the governance structure adapts to the dynamic demands of the data center environment while minimizing risks and enhancing efficiency. A clear and comprehensive governance framework is fundamental to maintaining an efficient and reliable data center. It defines roles, responsibilities, and processes, ensuring all changes and tasks are managed effectively and safely.

This minimizes disruptions, maximizes uptime, and fosters accountability. When rigorously implemented, the governance structure becomes a strategic asset that promotes compliance, resolves conflicts, and streamlines operations. By consistently reviewing and refining the framework, operators ensure it remains relevant and prepared to handle evolving challenges and opportunities within the industry. This ongoing process is an investment in operational excellence and long-term success.

The Cabinet Board Approval Process

The smooth operation of a data center, particularly one with multiple tenants, relies heavily on a clear and efficient work approval process. This process becomes even more critical when the work affects tenant infrastructure, requiring a structured approach that carefully balances tenant needs with the facility's operational requirements. The cabinet board approval process addresses this necessity by offering a framework for evaluating and approving work requests that involve tenant cabinets and nearby infrastructure.

The cabinet board lies at the heart of this process. It typically includes representatives from the Critical Facilities Management (CFM) team, IT operations, and other key areas, such as controls and reliability. The board's composition may vary depending on the size and needs of the data center. Still, its core principle remains the same: ensuring representation from all stakeholders affected by the proposed work. This multi-stakeholder setup fosters collaboration, prevents conflicts, and ensures decisions are informed by a complete understanding of the potential impacts on both the data center and its tenants.

Before reaching the cabinet board, work requests undergo an initial review by the CFM team. This preliminary assessment evaluates the scope of the work, its potential effects on other systems and tenants, and the resources required for execution. The team identifies potential hazards and conflicts, verifies compliance with safety regulations and operational guidelines, estimates potential downtime, and drafts a preliminary schedule. By thoroughly vetting requests in advance, the CFM team ensures the cabinet board's time is used efficiently and that decisions are based on well-prepared proposals.

Requests must be clearly and thoroughly documented. Each request should include a detailed description of the work, the reasons for undertaking it, a proposed timeline, and the necessary resources. Supporting materials, such as engineering diagrams, network layouts, and risk assessments, provide the cabinet board with the information necessary to make informed decisions. Complete and accurate documentation minimizes delays and keeps the review process moving smoothly.

After the initial review, the request is formally submitted to the cabinet board. Submissions typically involve a formal request form or an online system that tracks the process from submission through completion. This system ensures proper record-keeping and traceability. The cabinet board reviews the request, focusing on its impact on operational stability, tenant business continuity, and compliance with regulations. During this review, board members may ask clarifying questions or request additional information to address potential risks and prevent disruptions.

The board's decision-making process must be transparent and consistent. A clear set of criteria should guide the evaluation of each request. These criteria might include the urgency of the work, its potential impact on other systems and tenants, associated costs, and alignment with the data center's goals. Explicitly defining these criteria reduces bias and ensures fairness. Additionally, meeting minutes should document the rationale behind each decision. Making these records available to relevant stakeholders fosters accountability and builds trust among all parties involved.

The work approval process in a data center often involves multiple levels of authorization, depending on the complexity and potential impact of the request. Simple tasks, like rearranging cables within a tenant's cabinet, may only require approval from an engineer on the critical facilities team. However, more complex requests, such as upgrading Power Distribution Units (PDUs) or implementing major network changes, may require approval from a senior engineer or the CFM. Significant changes, like modifications to the building's HVAC system, would likely require input from executive management. This tiered structure ensures efficiency while maintaining accountability.

Consider a practical example. If a tenant requests to add a server rack to their cabinet space, the process is relatively straightforward.

The CFM team verifies available space and power capacity, and the Chief Engineer could approve the request due to its low impact. On the other hand, a tenant seeking a substantial network infrastructure upgrade involving the rerouting of fiber optic cables would present a more complex challenge. This type of work demands a detailed risk assessment, thorough engineering plans, and input from the IT department. It would require cabinet board review and possibly executive-level approval, depending on the scope and impact on overall operations.

Preventative maintenance offers another scenario. Suppose a critical piece of equipment in a tenant's area needs scheduled maintenance. While necessary for reliability, the affected tenants must be informed well in advance. The cabinet board would assess the maintenance plan to minimize disruptions, ensuring tenants have time to prepare for any downtime. The cabinet board also plays a vital role in resolving conflicts, whether between tenants or between tenants and the management team. Acting as a neutral forum, the board facilitates discussions and ensures disputes are resolved efficiently and fairly. This approach helps prevent minor disagreements from escalating into major issues, creating a harmonious and collaborative environment.

After work is approved, the critical facilities team creates a detailed work plan. This plan outlines each step, the timeline, and the required resources. It is shared with affected tenants to coordinate schedules and minimize disruptions. During implementation, the work is closely monitored to ensure adherence to the plan and to address any unexpected challenges quickly.

A post-implementation review evaluates the outcomes of the completed work. This review compares the results with the original request, assesses tenant disruptions, and identifies opportunities for improvement in future processes. This feedback loop is critical for refining the cabinet board approval process and ensuring continuous improvement. Thorough documentation remains a priority throughout the process, creating a detailed audit trail for compliance and identifying potential weaknesses for future enhancements.

The cabinet board approval process is more than just a procedural step. It builds a collaborative and efficient framework for managing the complex and interdependent ecosystem of a data center. By focusing on continuous improvement, this process enhances reliability, supports smooth operations, and ensures tenant satisfaction. Rather than a bureaucratic hurdle, the cabinet board is a cornerstone of effective governance and cooperative data center management.

Change Management and Control Procedures

Change management is the cornerstone of maintaining stability and preventing disruptions in a dynamic data center environment, and requires a well-structured and proactive change management process. This is more than just following procedures; it involves anticipating potential issues and minimizing their impact. A robust change management strategy is crucial for ensuring reliability and operational efficiency, particularly in multi-tenant environments where the actions of one party can have a significant impact on others.

Effective change management starts with thorough documentation. Every change, no matter how minor or significant it seems, must be properly recorded. This documentation serves several critical purposes. First, it creates a detailed audit trail, making it easy to trace each change back to its origin, the individuals involved, and the reasons for implementation. This is invaluable for troubleshooting, compliance audits, and demonstrating due diligence. Second, comprehensive records help identify patterns in change requests, which can reveal areas that need process improvements or preventive action. For instance, frequent requests for similar work may point to underlying systemic problems requiring attention.

Standardizing the documentation process ensures consistency and clarity. Each change request should include a unique identifier, a clear description of the proposed changes, the rationale behind the request, the people or teams involved, an implementation timeline, associated risks and mitigation strategies, required resources, and the status of the request throughout its lifecycle.

This standard format keeps everyone aligned and minimizes misunderstandings. A centralized digital platform, such as SharePoint or a dedicated change management system, is essential for efficient record-keeping. These platforms offer features such as version control, approval workflows, and reporting tools, making the process more streamlined and accessible to all relevant personnel.

The approval process for changes should follow a tiered structure, similar to the work approval process. Simple changes, like replacing a faulty server rack component or rearranging cables within a tenant's space, can often be approved by technical staff within the critical facilities team. These individuals are empowered to make decisions based on their expertise and established guidelines. However, their authority is limited to changes with minimal impact on other systems or tenants. Ongoing training and professional development are crucial for these personnel to ensure they remain knowledgeable and up-to-date with industry standards and regulations.

More complex changes that involve multiple systems, affect tenant operations, or require significant resource demands necessitate a stricter approval process. These changes might need to be reviewed and approved by senior engineers, the Chief Engineer, the CFM, or even the executive management team. This process requires a thorough evaluation of potential risks and impacts, often including consultations with affected parties.

Risk assessments are a critical part of the approval process for significant changes. These assessments must identify both immediate and long-term potential hazards and detail strategies to mitigate or eliminate them. The assessment should also evaluate how the change could impact overall data center operations and tenant business continuity. All findings from the risk assessment must be clearly documented and shared with stakeholders to ensure transparency and facilitate informed decision-making.

The implementation phase of change management demands precise planning and execution. A detailed plan must be prepared before initiating the change, outlining all steps, timelines, assigned responsibilities, and required resources. Contingency plans are crucial for addressing potential problems or unexpected complications. During implementation, progress should be carefully monitored to ensure it aligns with the plan. The individuals executing the changes must be properly trained and equipped to perform their tasks safely and efficiently. Communication is relevant here too. Maintaining regular communication during implementation is crucial to keep all stakeholders informed and to address any issues that arise promptly.

Stakeholders should be informed about changes that may impact them, ensuring transparency and collaboration. This minimizes surprises and encourages a shared understanding. Clear and concise communication through multiple channels, such as meetings, status updates, and formal notifications, is essential to prevent misunderstandings. Feedback mechanisms should allow stakeholders to express concerns or suggest improvements. Using a centralized communication system, like a dedicated project management tool, can streamline these efforts and provide a comprehensive record of all communications.

Evaluating the outcomes of the change through a post-implementation review is just as important. This review assesses the success of the change, highlights lessons learned, and identifies areas for improvement in future processes. The outcomes should be compared with the expectations outlined in the initial change request. This step also evaluates the impact on the data center's overall operations, tenant activities, and the efficiency of the change management process. Challenges encountered during the implementation should be thoroughly analyzed to identify areas where adjustments are needed. Stakeholder feedback plays a key role in this review, often uncovering hidden issues or potential improvements that might otherwise go unnoticed.

Implementing new technologies or significant upgrades requires a specialized approach. These changes often involve integrating new systems with existing infrastructure, a process that demands expertise in multiple disciplines. A thorough testing phase is critical to ensure compatibility and identify potential conflicts before deployment. Testing should mimic real-world scenarios to expose any unforeseen problems or bottlenecks. Comprehensive staff training is equally important. Training should combine theoretical knowledge with hands-on experience to ensure the team fully understands the new technology and its operation. For substantial software updates, rollback plans are essential to enable a quick recovery in the event of unexpected issues that arise after deployment.

Training staff on this policy ensures they understand and follow its guidelines, promoting consistency and adherence across the organization.

Fostering a proactive culture around change management strengthens its effectiveness. Encouraging staff to report potential issues, identify areas for improvement, and view change requests as opportunities helps build a collaborative and forward-thinking environment. Regular audits and reviews of the change management process identify weaknesses and areas for improvement, ensuring the process remains effective and efficient.

Investing in training at all levels enhances the success of change management. A well-defined and consistently applied process is more than a procedural requirement—it is critical to ensuring the reliable and efficient operation of a modern data center. By embedding robust change management into the organization's culture, data centers can maintain stability and security, benefiting both operators and tenants.

Tracking and Reporting on Work Orders

Tracking and reporting work orders effectively are essential for a data center to run smoothly. This process not only keeps teams responsible for completed tasks but also provides transparency into ongoing projects. Clear visibility helps manage resources and address potential delays proactively. A dependable work order tracking system also creates a solid audit trail, which is vital for internal review and external compliance. This section emphasizes building and maintaining such a system, highlighting the advantages of using dedicated project management tools.

Establishing a standardized format is the first step in creating an efficient work order tracking system. Each work order should have a unique identifier for easy retrieval and reference. Depending on the size of operations and the software in use, this identifier could be a straightforward numerical sequence or a more complex alphanumeric code. The work order itself must include a detailed description of the task. This description should specify the exact location within the data center, the equipment involved, and the desired outcome. Ambiguity must be avoided at all costs. Clear instructions reduce the chances of misinterpretations that could lead to delays or mistakes. For more complex tasks, attaching diagrams or sketches can be incredibly helpful in providing additional clarity.

Prioritization is another essential element of a work order. Each task should be classified by its priority level, ranging from "critical" (requiring immediate action) to "low" (which can be scheduled for later). This classification not only indicates urgency but also considers the potential impact on tenants' operations. A task that might seem minor could have significant downstream effects on tenant services, justifying a higher priority. Assigning priority levels also guides resource allocation, ensuring that high-priority tasks receive prompt attention from the most qualified personnel.

Work orders must also specify the assigned personnel responsible for the task, including their contact information. Clear accountability is essential for streamlining communication and preventing delays caused by uncertainty. For larger projects that require input from multiple teams, assigning a project manager to oversee coordination and progress is critical. The project manager ensures that all teams work cohesively, deadlines are met, and communication is seamless throughout the project. Timelines are equally important and should be realistic. A work order shouldn't only include a completion date—it should also break the task into smaller milestones, each with its own deadline. This approach makes it easier to monitor progress and spot potential delays early.

Timelines should factor in more than just the task itself; considerations such as equipment availability, team schedules, and the likelihood of unexpected complications must also be taken into account. These milestones should be aligned with the overall project deadline, with the tracking system set to flag any deviations from the plan for immediate attention.

The actual work performed should be documented. Documentation includes not only a summary of the completed tasks but also details about any unexpected issues that arose, the corrective actions taken, and the materials used. Maintaining this level of detail creates a valuable record for future troubleshooting, preventative maintenance, and performance analysis. It also helps evaluate the process's efficiency and highlights areas for improvement. Additionally, this information contributes to the data center's operational history, building a knowledge base that serves as a reference for future work.

The final approval section of the work order is just as important. This section requires sign-off from both the personnel who performed the work and the supervisor or manager overseeing the process. The approval should include details like the completion date and time, confirmation that the work met the required specifications, and notes on any deviations or adjustments made.

This dual-approval process ensures accountability and verifies the quality of the work that has been completed. Digital signatures, timestamps, and photographic or video evidence add an extra layer of integrity to the process by creating a verifiable record of completion and inspection.

Generating regular reports is another essential aspect of work order management. These reports, ideally produced automatically by the system, provide insights into the efficiency of the data center's operations. Key metrics such as the average time taken to complete work orders, the number of outstanding tasks, completion rates, and the frequency of specific types of requests can be tracked. This data helps identify bottlenecks, improve processes, and optimize resource allocation. Managers should have the ability to customize reports, filtering data by priority levels, locations, or assigned personnel. This flexibility allows for targeted analysis of specific areas or teams, making the process more effective and actionable.

Utilizing project management software provides numerous benefits for managing work orders. Such platforms provide centralized storage for all work orders, ensuring easy access and reducing the risk of losing critical documents. Features like automated notifications, real-time progress tracking, and automated report generation streamline the entire process. Many platforms also support collaboration, enabling multiple team members to work on the same work order simultaneously, thereby enhancing teamwork and communication. Risk management features in these platforms further add value by proactively identifying and addressing potential issues, minimizing delays and disruptions.

Choosing the right project management software depends on several factors, including the size of the data center, the complexity of its operations, and budget considerations. Larger and more complex data centers may need software with advanced features, while smaller facilities might prefer simpler, cost-effective solutions. Evaluating options based on scalability, usability, and integration with existing systems is essential before making a decision. Taking advantage of trial periods or vendor demonstrations can provide a better understanding of each platform's suitability. Seamless integration with the current IT infrastructure is crucial to prevent additional workflow complexities and ensure smooth adoption.

Ensuring staff are properly trained to use the work order system is essential for maximizing its effectiveness. Training should cover all aspects of the system, from submitting and managing work orders to generating reports and monitoring progress. Ongoing refresher training helps staff maintain their skills, ensuring the system continues to deliver its full benefits. Training should include all relevant personnel, such as technicians, engineers, managers, and even tenants who may need to submit work requests. A workforce that is confident and proficient in using the system significantly contributes to the overall efficiency of the work order management process, thereby reducing errors and improving outcomes.

The implementation of a robust work order tracking and reporting system, ideally supported by reliable project management software, is crucial for efficient data center operations. This system promotes accountability, enhances transparency, and enables proactive management of resources and tasks. Streamlining internal processes and enhancing communication with tenants fosters a positive and productive working environment. The system also supports better collaboration between teams, ensuring tasks are completed on time and to the required standards.

Investing in a well-structured system and thorough staff training is a key step in reducing disruptions and ensuring maximum uptime for critical infrastructure. It not only optimizes internal operations but also strengthens relationships with tenants by promptly and efficiently addressing their needs. Regularly reviewing and refining the system based on performance data ensures it remains effective and adapts to evolving operational demands. A robust system, combined with continuous improvement and training, lays the foundation for the long-term success and reliability of the data center's operations.

Ensuring Compliance with Regulatory Frameworks

The complexity of operating a modern data center requires strict adherence to a wide range of regulations. These facilities are not just subject to basic building codes—they must also comply with data privacy laws, environmental regulations, and industry-specific standards. Meeting these requirements is essential to maintaining operational integrity and avoiding costly penalties. This section outlines the critical steps in creating and maintaining a compliance program that keeps a data center aligned with all relevant legal and regulatory frameworks.

Identifying all applicable regulations is the foundation of an effective compliance program. This task can be challenging, as data centers often operate under multiple jurisdictions and regulatory bodies. Local building codes and fire safety standards establish minimum requirements for construction, safety systems, and emergency procedures. Beyond these, national laws governing data privacy, such as GDPR in Europe, CCPA in California, and PIPEDA in Canada, impose strict rules on data storage, access, and transfer. Industry-specific standards, including those from the Uptime Institute, TIA-942, and ISO 27001, serve as benchmarks for best practices in data center design, construction, and operation. Conducting a comprehensive regulatory audit, ideally with the assistance of legal or compliance experts, is crucial for identifying all applicable rules and any gaps in current practices.

Creating a detailed compliance plan is the next step. This plan should go beyond a simple checklist and act as a roadmap that outlines the necessary steps to meet regulatory requirements. It should include specific procedures, clearly assigned responsibilities, timelines, and measurable metrics for tracking compliance. For instance, a data privacy plan might specify encryption protocols, access controls, data retention policies, and breach response procedures. Similarly, environmental compliance plans should cover energy efficiency goals, waste management strategies, and compliance with emissions standards. Connecting these plans with existing work approval processes guarantees that all activities are aligned with regulatory requirements. Any proposed work that impacts compliance must undergo a thorough review process before approval.

A dedicated team is essential for enforcing and monitoring compliance. This team could include internal experts, external consultants, or a combination of both. Their role includes conducting regular audits, providing staff with training on regulatory requirements, and keeping detailed records of all compliance-related activities. These records not only serve as proof of compliance but also provide valuable insights for ongoing improvement. Regular reports to management on the compliance status help identify potential issues before they become major problems. The frequency of these reports should reflect the complexity of the regulatory landscape and the data center's specific operations.

Maintaining compliance in a modern data center is not a one-time effort. It requires constant vigilance, regular updates to the compliance plan, and proactive efforts to address evolving regulations and operational changes. By taking a structured and well-documented approach, data centers can ensure they remain aligned with legal and regulatory requirements while fostering a culture of accountability and continuous improvement.

Training is a critical component of maintaining compliance in data center operations. Everyone involved, from technicians and engineers to managers and executives, needs appropriate training in relevant regulations. This training should be tailored to each individual's role, ensuring everyone understands their specific responsibilities under the law. It should also cover the serious consequences of non-compliance, such as fines, legal action, and damage to the organization's reputation. To keep knowledge fresh and up-to-date, regular refresher sessions must be conducted, especially when regulations change. A workforce equipped with a strong understanding of compliance lays the foundation for an effective compliance program.

Technological tools play a crucial role in supporting compliance efforts. Data Center Infrastructure Management (DCIM) software can track and monitor parameters essential to meeting regulatory requirements, such as energy usage for environmental compliance. Security Information and Event Management (SIEM) systems enable real-time monitoring of security incidents, ensuring compliance with data privacy laws. These systems are invaluable in creating an evidence trail for audits, helping demonstrate compliance with minimal effort. Access control systems and logging mechanisms further enhance security, effectively protecting sensitive data and ensuring compliance with privacy regulations.

Detailed records of compliance activities must be maintained to demonstrate regulatory adherence. These records should include permits, licenses, certifications, and documentation of inspections, audits, training sessions, and corrective actions. Using a centralized document management system, like SharePoint, makes organizing and retrieving these records more efficient. Regularly reviewing and updating documentation ensures it reflects the latest regulations and operational practices.

This level of organization is vital for facing external audits or legal actions with confidence.

Keeping up with the constantly changing regulatory landscape requires ongoing effort. Regulations evolve continuously, making it crucial to stay informed about updates and new requirements. Monitoring industry publications, attending relevant events, and maintaining open communication with regulatory bodies are effective ways to stay ahead. Regularly reviewing the compliance plan ensures it aligns with current laws and the data center's operational changes. Taking a proactive approach to compliance rather than reacting to issues as they arise reduces risks and helps sustain operational efficiency. Anticipating future requirements and smoothly integrating necessary updates into daily practices ensures a seamless transition whenever changes occur.

Clear accountability within the compliance structure is essential. Assigning responsibility to a dedicated compliance officer or team ensures the program receives proper attention and resources. This team should report directly to senior management to highlight the importance of compliance in the organization. Their duties include developing and updating the compliance plan, conducting audits, training staff, and maintaining thorough records. Clearly defined roles and reporting lines prevent confusion, encourage accountability, and allow for the efficient resolution of any compliance issues.

Integrating compliance into the data center's governance structure strengthens the overall operational strategy. Compliance should not operate in isolation but should be woven into every aspect of the data center, from design and planning to construction and maintenance. A well-rounded governance framework encourages collaboration among stakeholders, including management, staff, tenants, and external partners. This integrated approach avoids duplication of effort and aligns operational efficiency with regulatory standards. Regular reviews of the governance framework and compliance program ensure they stay effective and adaptable. Periodic external audits provide an objective assessment of compliance status and offer valuable feedback to refine governance and compliance processes.

Data Center Infrastructure Management (DCIM) Tools

Data Center Infrastructure Management (DCIM) tools are now a must-have in any well-run data center. These tools provide a comprehensive suite of features that enhance efficiency, reduce costs, and facilitate informed decision-making. DCIM consolidates everything into a single, central platform, where teams can monitor, manage, and analyze various aspects of the data center. This setup enables proactive control over operations, helping to prevent problems before they escalate. A major strength of DCIM is its ability to give a clear, single view of the entire facility by pulling data from different systems and showing it in a way that's easy to understand. With this setup, data center staff can quickly check the status of key systems, identify issues early, and take action before problems arise.

Improving day-to-day efficiency is one of the biggest advantages of using DCIM tools. These systems can automate common tasks, such as planning for future capacity, managing power and cooling, and tracking the utilization of rack space. With these tasks handled automatically, IT staff can focus on larger, more important projects. For example, smart capacity planning features can analyze past trends and future needs, helping teams plan expansions before reaching limits that could cause service slowdowns. DCIM also helps manage power and cooling more effectively by ensuring resources are used efficiently. This reduces wasted energy and lowers operating costs. Tools that handle space planning help avoid unused rack areas and boost density, which is especially helpful when trying to grow without using more floor space.

Running a data center can be expensive, but DCIM tools help keep costs under control. By automating routine tasks and using resources more efficiently, the need for hands-on fixes goes down. This results in significant savings over time. DCIM makes it easier to identify energy waste by analyzing how power is used and highlighting where small changes can have a significant impact. Sometimes that just means adjusting temperature settings or fine-tuning how power flows through the system. The tools can also identify underused or outdated equipment, helping teams determine what to retire or consolidate. That kind of insight can avoid the need to buy more hardware. Catching issues early also helps avoid expensive downtime and last-minute repairs.

Better decision-making is another significant benefit of DCIM. The platform gives a clear picture of how everything is running. This makes it easier to plan upgrades, shift resources, and schedule maintenance with confidence. The built-in reporting features let you see key performance numbers at a glance and help track how things are improving. Over time, the data builds up and reveals patterns that aren't obvious just by watching one system at a time. That kind of long-term view helps spot problems before they start. Reports and dashboards transform numbers into visuals that are easy to understand, making it easier to share progress and identify areas that require attention.

The market offers a wide range of DCIM software tools, each with its own unique strengths and limitations. Some of the most well-known options include Schneider Electric's StruxureWare Data Center Expert, CA Technologies' Spectrum, and Nlyte Software. These tools usually come with features such as capacity planning, power and cooling management, space tracking, and change management. However, the features available can differ from one vendor to another and may also vary by software version. Choosing the right tool means looking closely at how well it matches your data center's unique needs.

Schneider Electric's StruxureWare Data Center Expert stands out for its all-around capabilities. It covers nearly every aspect of data center infrastructure and offers advanced tools for predictive maintenance and capacity planning. It helps manage both power and cooling systems more efficiently, reducing energy consumption while enhancing performance. The software also includes detailed reporting tools that provide insights into system health and performance, enabling teams to identify trends and respond promptly to potential issues.

CA Technologies' Spectrum focuses more on managing IT infrastructure. While it can help with physical systems to some extent, its primary strength lies in providing a unified view of servers, storage devices, and network equipment. This makes it a good fit for companies seeking to manage both virtual and physical systems from a single dashboard. It consolidates everything into one place, enabling IT teams to stay on top of system performance and respond quickly to issues.

Nlyte Software is another strong DCIM solution that places a significant focus on capacity planning and managing infrastructure changes. It enables teams to map out their full setup in detail, which helps them forecast power and space needs more accurately. Its change management tools help ensure that any updates or changes are properly recorded and approved, reducing the risk of mistakes or disruptions.

Picking the right DCIM tool depends on many factors—how big the data center is, how complex the setup is, what the company needs from the system, and how much they can spend. A detailed review of each option is essential. This involves comparing the available features, assessing the system's scalability, and determining how well it integrates into the current environment. When done correctly, a DCIM solution can make a big difference in how smoothly the data center runs, how much it costs to operate, and how informed decision-making becomes.

Several advanced DCIM platforms go beyond the basics, offering features such as automated workflows, predictive analytics, and seamless integration with other tools. Automated workflows reduce manual work by handling routine tasks in the background, freeing up resources for more complex tasks. This saves time and lets the IT team focus on more important goals. Predictive analytics, powered by machine learning, helps catch problems early by looking at past patterns and alerting the team before things break. Integration with other systems, such as building management platforms (BMS) and IT service management tools (ITSM), helps create a comprehensive view of the entire operation, from the floor to the cloud.

A successful DCIM setup doesn't just depend on the software—it needs proper planning. Every project should begin with a clear understanding of what the company wants to achieve. From there, it's important to choose a system that works well with existing tools, and then create a plan that covers setup, timelines, and roles. Everyone involved should know what's expected of them, and any risks or roadblocks should be noted and planned for in advance. This helps avoid delays and makes the rollout smoother. Maintaining the system after launch is just as important as the setup. Software updates, regular checks, and ongoing support are all essential to maintaining the system's optimal performance. A strong support contract with the software vendor can also help by offering quick help when problems come up.

Training should not stop after the first round—ongoing sessions make sure staff stay up to date and can use all the features the system has to offer.

DCIM tools have become essential for managing data centers effectively. They help monitor and control everything from power to space, leading to improved resource utilization, lower costs, and more informed decision-making. The right DCIM tool, along with solid planning, regular updates, and well-trained staff, can transform how a data center is run. The benefits go beyond money and efficiency—they include stronger operations, quicker responses to problems, and a better understanding of the entire environment. Taking a smart and detailed approach to choosing and setting up a DCIM system is the best way to ensure the investment pays off.

Building Management Systems (BMS)

Building Management Systems (BMS) are essential for overseeing the overall environment of a modern data center. While DCIM tools mainly focus on the IT side—such as servers, storage, and network equipment—BMS encompasses the wider building systems. This includes HVAC, power distribution, fire safety, physical security, and other critical components that keep the facility operational. When BMS is properly integrated with DCIM and other management tools, it becomes much easier to enhance efficiency, reduce energy consumption, and prevent unplanned outages.

A BMS works as a central control system that keeps track of different parts of the building using sensors, controllers, and actuators. These tools gather data around the clock, monitoring conditions like temperature, humidity, airflow, and power usage. The system sends that data to a central platform where it's reviewed and used to make decisions. Based on the data, the BMS can automatically adjust building systems. For example, if the temperature rises above the target level, the BMS can boost cooling efforts to bring it back down. This real-time response helps protect equipment and maintain a stable environment.

One of the biggest advantages of using a BMS in a data center is its ability to improve visibility and early detection. Since it monitors multiple parts of the facility simultaneously, it can quickly identify when something isn't working as expected. This helps teams address small issues before they escalate into larger problems. If an HVAC unit starts acting up, the BMS can send an alert before the temperature climbs high enough to cause damage. If power use suddenly spikes, the BMS can flag it, giving the team time to step in before it leads to a failure or outage.

Another important role of a BMS is to control various systems, enabling them to run more efficiently. It can adjust heating, cooling, humidity, and lighting based on real-time conditions to maintain the right environment without wasting energy. Over time, this can result in significant cost savings. Many newer BMS setups utilize smart algorithms and predictive tools to determine the most efficient way to utilize energy, based on real-time data and anticipated changes. This type of smart control enables the facility to remain efficient without requiring constant manual input.

Keeping energy use in check is a top priority for any data center, and BMS plays a major part in that. The system tracks how much energy is used, where it's used, and when. That information helps teams identify areas where they can make savings. They might adjust cooling setpoints, create smarter HVAC schedules, or even use demand-response programs to reduce load during peak hours. Since cooling is one of the biggest energy expenses in a data center, any improvement there can make a real difference. One way the BMS helps is by utilizing free cooling when outdoor temperatures are sufficiently low. Instead of relying only on chillers, the system pulls in cool outside air to maintain temperature, cutting down on mechanical cooling costs.

The BMS can also improve how CRAC (Computer Room Air Conditioning) and CRAH (Computer Room Air Handling) units operate. It adjusts their speed and power output based on how much cooling is needed in real time. This means the system only uses the energy it truly needs, avoiding waste while still protecting equipment from overheating.

Beyond energy use, BMS adds value by improving safety and security. It works with fire systems to monitor for alarms and activate suppression systems when needed. It can also integrate with security features such as access control, video cameras, and motion detectors. This makes the building safer and ensures that only authorized people can enter sensitive areas. Together, these features make the data center more secure and reliable. When combined with DCIM and other tools, BMS gives operators a full picture of what's going on inside the data center. It helps prevent problems, supports better energy use, and keeps people and equipment safe. With smart planning and good integration, BMS becomes more than just a control system—it becomes a central part of how a data center runs smoothly every day.

The integration of Building Management Systems (BMS) with other data center tools, such as DCIM platforms, is crucial for establishing a comprehensive, unified management solution. This kind of integration allows both systems to share data, giving operators a full picture of the data center's performance and environment. With both tools working together, teams can make better decisions, stay ahead of problems, and enhance the daily operation of the facility.

This type of system-to-system collaboration enables smart, automated responses. For example, if the BMS notices that the temperature is climbing past a safe limit, it can send that information to the DCIM platform. From there, the DCIM system can trigger specific actions, such as increasing the cooling output or shifting power loads to mitigate risk. This level of coordination helps prevent service disruptions and protects equipment from damage.

Some large data centers already use advanced BMS setups that show just how valuable these systems can be. These platforms don't just control HVAC and lighting—they also use data analytics to predict issues before they cause problems. With the help of sensors and smart software, BMS systems can alert staff about possible failures before they happen. For example, if a cooling unit starts showing unusual performance trends, the system can flag it for maintenance before it fails, reducing downtime and repair costs.

BMS also plays an important role when it comes to meeting regulatory requirements. Data centers must adhere to strict guidelines regarding energy consumption and environmental impact. The data collected by a BMS makes it easier to prove compliance. Reports and logs generated by the system track energy usage, system performance, and the amount of cooling delivered. This level of detail can support audits and help meet strict efficiency goals.

Before installing a BMS, data center operators need to plan carefully. Every facility has different needs, and the planning stage should include a comprehensive review of the systems that will connect to the BMS. Teams should select the appropriate hardware and software, define the system's capabilities, and outline its intended use. Just as important is ensuring that all staff members are properly trained to use and maintain the system. A well-informed team is key to getting the most out of the investment. Once the system is up and running, keeping it maintained is critical. This involves applying software updates, inspecting the hardware for issues, and performing regular maintenance to ensure everything operates as intended.

Ongoing support from the vendor can help resolve problems quickly and ensure the system remains reliable. When a BMS is looked after properly, it can deliver strong performance for many years, making it a solid long-term asset.

As more data centers combine BMS with other tools, the benefits become even clearer. Integration leads to smoother operations, lower costs, and fewer unexpected failures. What was once considered a bonus has become a standard part of smart data center management. The key is to set clear goals from the outset, select systems that work well together, and ensure the individuals running them are fully trained and supported.

Today's advanced BMS platforms do more than just monitor; they help teams plan, predict, and adjust. The data they collect can be used to cut energy waste, improve uptime, and keep the entire data center safer and more efficient. This also sets the stage for future growth and innovation, particularly as demands on data centers continue to increase. Predictive maintenance, smart automation, and better visibility across systems all come together to support strong, steady performance.

By integrating BMS and DCIM systems with a clear strategy, data center operators can create an infrastructure prepared for any situation. Combining solid planning, the right technology, and well-trained staff results in better outcomes, fewer surprises, and long-term success. The ability to anticipate issues, cut unnecessary costs, and maintain reliable performance makes BMS an essential part of any modern data center.

Leveraging SharePoint for Collaboration and Documentation

SharePoint does much more than store files; it has become a reliable platform for managing the numerous components of data center operations. Its ability to bring documents, communication, and project updates into one place makes it a key tool during the transition from construction to a fully operational facility. By organizing everything in a structured way, SharePoint makes important information easy to find and use, which helps reduce mistakes and keeps teams working together smoothly.

One of the biggest advantages of using SharePoint is its ability to simplify documentation. Instead of digging through emails, multiple folders, or outdated paper copies, teams can store everything in one searchable location.

This includes blueprints, technical specs, operation manuals, service logs, and more. With everyone working from the same source, there's less confusion and wasted time. For example, a separate SharePoint site can be created for each major system— like HVAC, power, or security. Each one can hold detailed diagrams, maintenance records, parts lists, and schedules. This setup enables technicians to obtain the exact information they need when handling repairs or checks, thereby minimizing downtime and enhancing productivity.

Version control is another key strength in a data center, where updates occur frequently and accuracy is crucial; it's essential to track every change. SharePoint handles this by logging each edit automatically and allowing users to roll back to earlier versions if needed. This feature is especially helpful during troubleshooting or system reviews. For instance, if a new software update causes performance issues on a server, admins can quickly pull up previous versions of the system settings to find out what changed. This shortens the time needed to resolve the issue and helps prevent further disruptions, resulting in improved uptime and reduced costs associated with system failures.

Teamwork also gets a major boost from SharePoint. With real-time editing, multiple team members can work on the same document at once, which cuts down on back-and-forth emails and version mix-ups. Whether it's the commissioning team posting test results or the construction crew updating their schedule, everyone can see updates as they happen. This shared space builds trust and keeps everyone aligned, making project coordination more effective. Instead of working in silos, teams stay connected and informed. The platform's built-in workflow tools take things a step further. Custom workflows can manage the approval process for critical updates, ensuring that every change is reviewed before being implemented. This reduces risk and adds a layer of control over updates to systems and layouts.

For example, when someone proposes a change to the data center floor plan, SharePoint can automatically route the request to architects, engineers, and facility managers for feedback and approval. This ensures that nothing is missed and that safety rules are always followed. By reducing delays and manual steps, the process becomes smoother and more dependable.

Each of these features—centralized documentation, version tracking, collaboration, and workflows—makes SharePoint more than just a document library. It becomes a vital tool that helps manage the complexity of data center operations. When everyone uses the same platform, information flows more freely, decisions happen faster, and the whole operation runs more smoothly.

Planning also becomes easier. With records of every step, every change, and every team involved, it's possible to look back at what worked and what didn't. This kind of insight helps with future builds, upgrades, and ongoing maintenance. It also ensures that teams don't repeat the same mistakes, which saves time and money in the long run. Bringing SharePoint into the data center environment isn't just about having a place to store documents—it's about building a system that supports clear communication, fast action, and strong accountability. With everything in one place and the right tools to manage it all, SharePoint helps data centers stay on track and handle complex projects with confidence and control.

The use of SharePoint for training and knowledge sharing offers major benefits in data center operations. It provides teams with a central location to store training materials, videos, user guides, and other valuable resources. With everything in one spot, staff can easily find the information they need to do their jobs correctly. This setup not only makes training more consistent but also lowers the chances of mistakes. New hires, for instance, can go through onboarding content that covers standard operating procedures, emergency steps, and safety rules. At the same time, experienced technicians can pull up maintenance instructions, repair tips, or technical specs without having to track down documents from different systems or team members.

The project management tools built into SharePoint offer even more value. Features like task assignments, deadline tracking, and issue reporting help project leaders keep a close eye on every stage of a job. With this kind of visibility, it's easier to spot problems early and make quick adjustments. That kind of control helps avoid delays and keeps costs under control. When building a new data center, for example, every phase of the project can be tracked using SharePoint. Each task can be assigned to a person or a team, while the system tracks progress and highlights any tasks that are falling behind. Managers can then shift resources or tweak the schedule as needed to keep things moving.

The ability to connect SharePoint with other tools makes it even more effective. When linked with DCIM systems, for example, SharePoint can give users a fuller picture of what's happening across both the IT and facility layers of the data center. This helps teams work together more effectively, improves decision-making, and enhances reporting. Picture a situation where the DCIM system picks up on a rise in server room temperature. By syncing with SharePoint, an alert can be sent straight to the right people, giving them a chance to fix the issue before any damage is done. That kind of rapid response keeps systems stable and prevents more significant problems.

It's not enough to just install the software and hope it works. A proper rollout involves determining who's responsible for what, establishing a logical folder structure, and establishing clear guidelines for document naming and storage. This makes everything easier to find and keeps the system clean. On top of that, all users need to be trained—not just on the basics like uploading files or finding folders, but also on more advanced tools like setting up workflows or creating custom alerts. When people understand how to utilize these features, SharePoint becomes an integral part of their everyday work, rather than an additional task to manage.

Security must also be taken seriously, especially in a data center setting. Only authorized staff should be able to view or change sensitive information. This requires setting strict access controls and regularly reviewing them to ensure they remain up to date. Security audits and penetration tests should also be done on a routine basis to catch weak spots. Strong security includes using two-factor authentication, encrypting sensitive files, and keeping the software fully updated. These steps protect data from both internal and external threats, ensuring the system remains secure at all times.

Cost is another factor that can't be ignored. Organizations need to think through the full price of SharePoint—not just the license fee, but also the cost of servers (if hosting on-prem), the time spent on training, and the ongoing support needed to keep the platform working well. Running a cost-benefit analysis helps weigh these expenses against the benefits, such as faster processes, fewer errors, and improved teamwork. When done correctly, the benefits far outweigh the investment; however, thorough planning is necessary to make that happen. When used strategically, SharePoint becomes more than just a place to store documents. It turns into a powerful tool that supports nearly every part of the data center—helping manage change, share knowledge, and keep everything running smoothly.

From training and documentation to communication and task management, SharePoint impacts every aspect of daily operations. Its automation features cut down manual work, while its real-time collaboration tools bring teams closer together.

Success comes from planning carefully, training users properly, and maintaining the system over time. With those pieces in place, SharePoint becomes a trusted part of the data center's daily workflow—one that helps people do their jobs faster, better, and more securely. The result is a smarter, more connected, and more resilient operation ready to take on whatever comes next.

Remote Monitoring and Management Technologies

Remote monitoring and management technologies are crucial for maintaining the smooth and efficient operation of modern data centers. These tools enable teams to monitor equipment, identify issues, and resolve problems remotely, without the need for on-site presence. This saves time, lowers costs, and helps prevent outages. When used correctly, these technologies enhance reliability and enable prompt responses to technical issues that might otherwise result in major disruptions.

One of the key tools in this area is remote power management. This enables teams to control and monitor devices such as Power Distribution Units (PDUs) and Uninterruptible Power Supplies (UPSs) from a central location. Advanced PDUs even allow you to control power at the outlet level. That means if a piece of equipment locks up or needs a reset, a technician can remotely power cycle just that one outlet. This is a big help during troubleshooting and also supports preventative maintenance. In addition, these systems provide detailed data on power use, voltage, and load levels. When something unusual happens—like a sudden spike in voltage or a drop in power—alerts can be sent out, so staff can jump in before the problem grows. Early warnings can make the difference between a quick fix and a costly outage.

Remote reboot capability is another time-saver. If a server crashes or software freezes, IT staff can restart it from a different location without waiting to be physically on-site. For example, if a glitch causes a critical server to hang, the administrator can initiate a remote reboot within minutes. This speeds up recovery time, improves uptime, and reduces the need to send someone into the facility just to push a button or flip a switch.

In addition to power systems, environmental monitoring is just as important. Sensors placed throughout the data center measure temperature, airflow, humidity, and other environmental factors in real time. These readings are sent to a central system that tracks performance and sends alerts when conditions fall outside the specified range. If, for instance, the temperature inside a rack starts rising too quickly, it could be a sign that a cooling unit has failed. The monitoring system can flag this immediately, allowing the team to respond before servers overheat. This keeps equipment safe and reduces the risk of service interruptions.

Cameras are often used in conjunction with environmental sensors. Placing visual monitoring systems around the data center gives teams an extra layer of oversight. Video feeds paired with environmental data help operators get a clearer picture of what's happening. Whether it's a blocked airflow vent, a tripped breaker, or unauthorized access to a sensitive area, remote visibility means faster responses and better control over the facility.

Remote infrastructure management extends beyond power and climate control. It also includes the ability to access and control core equipment, such as servers, switches, and storage systems. With the right tools, technicians can log into device dashboards, apply software patches, adjust configurations, and even monitor performance metrics. This means issues can be caught early and resolved quickly. For instance, if a top-of-rack switch shows signs of slowing down, remote access allows the team to run diagnostics, fine-tune settings, or roll out a firmware update— without setting foot in the data center.

The ability to manage all these systems from a single location makes daily operations smoother and reduces the burden on staff. Teams don't have to wait around for problems to escalate, and they don't need to send someone in person for every fix. Instead, they can solve issues the moment they're detected, and many problems can be avoided entirely through ongoing monitoring.

Remote monitoring also improves long-term planning. By collecting data on how systems perform over time, administrators can spot trends and plan upgrades or replacements before failure occurs. Power usage patterns, temperature shifts, and equipment performance logs help shape smarter decisions about maintenance, capacity, and future builds.

These technologies give data center managers peace of mind, knowing they can monitor every system and react quickly from anywhere. As data centers expand and change, remote monitoring and management will remain vital in ensuring uptime, security, and efficiency—all while controlling costs and response times.

Network monitoring tools are key to keeping a data center's network healthy and performing as expected. These tools track real-time data, such as bandwidth usage, traffic flow, and latency levels. With more advanced systems, administrators can pinpoint slowdowns and other performance issues before they cause real problems. Most platforms offer visual dashboards that display the overall status of the network, enabling teams to identify and resolve issues promptly. When predictive analytics is added to the mix, network monitoring becomes even more useful. By reviewing past data, these tools can identify trends and predict potential outages before they occur. That means maintenance can be scheduled in advance and downtime can be avoided.

Remote hands services give data centers a way to respond physically when remote troubleshooting isn't enough. These services provide trained on-site technicians who can follow instructions from remote administrators. Whether it's swapping out a failed drive or connecting a new switch, remote hands staff can step in without needing a visit from the client's own team. This saves time and money, especially for businesses managing data centers remotely. Having skilled help on demand ensures tasks are completed quickly, without delays due to travel or scheduling.

Choosing the right remote monitoring and management setup depends on the size of the data center, the complexity of its systems, and what the business truly needs. Cost is also a factor. Starting with the most important monitoring tools and gradually expanding from there is often the most effective approach. It's also important to focus on security. Any remote system must have strong access controls to block unauthorized users and keep the entire environment safe.

Having the right tools is crucial, but having a strong strategy is equally important. Every team should have clear Standard Operating Procedures (SOPs) in place for handling problems remotely. These SOPs need to cover everything from diagnostics to escalation plans. Training should also be ongoing. Team members must be familiar with the tools and confident in using them to solve problems.

Solid communication between remote teams and on-site staff is another critical piece. When something goes wrong, both sides need to work together quickly. A centralized ticketing system helps keep things organized, ensuring each issue is tracked and handled properly. That way, no problem slips through the cracks, and response times stay short even during emergencies.

Planning for remote monitoring and management isn't just about tools—it's about designing a system that works under pressure and can grow over time. Before setting anything up, a full needs assessment should be done. This review should examine current requirements and also consider future needs of the data center. Systems must be built with redundancy to avoid single points of failure and with scalability to support future growth and expansion.

Ongoing upkeep is essential. Software updates, hardware checks, and regular security audits must be part of the routine. If these tasks are skipped, performance can suffer, or worse, security can be compromised. By staying on top of these details, teams make sure the monitoring systems remain reliable and secure.

The impact of remote monitoring and management is significant. These tools reduce downtime by detecting issues early and resolving them quickly. They lower operating costs by reducing the need for site visits. They also enhance how fast teams can respond, making the entire operation more agile. Better oversight also means improved security. When administrators have full visibility into what's happening across the facility, they can quickly lock things down or take corrective measures. Finally, these systems make expansion easier. Adding new servers, devices, or entire sites is simpler when a remote management system is already in place.

Remote monitoring and management aren't just useful extras anymore—they are a core part of modern data center operations. Without them, it's harder to maintain availability, meet performance goals, or control costs. With them, facilities become more efficient, more secure, and more flexible. The tools and strategies you choose will have a big impact on how your data center performs today and how well it adapts tomorrow.

Technology in this space continues to advance, offering new ways to enhance reliability and performance. Staying updated on the latest options is important for any data center manager. Those who keep learning and adapting are the ones who stay ahead of problems, reduce risk, and make the most of their infrastructure.

Predictive Analytics for Data Center Optimization

Predictive analytics utilizes a combination of historical data, real-time sensor readings, and intelligent algorithms to identify potential issues and performance trends in the data center. This proactive method represents a significant step forward from the older, reactive approach, where teams only address problems after something goes wrong. Waiting for failures leads to costly downtime and disruption. With predictive analytics, managers can identify problems early, take preventive action, and utilize resources more effectively. The result is fewer breakdowns, improved efficiency, and stronger overall reliability—along with significant cost savings.

At the heart of predictive analytics is solid data collection. To make accurate forecasts, the system needs a wide range of information from across the data center. This includes power usage statistics, environmental readings such as temperature and humidity, network traffic reports, and IT performance data. The more details it has, the better it works. For example, reviewing power usage from past months can reveal trends, such as seasonal demand peaks. Knowing this ahead of time helps with planning, so power isn't stretched too thin during busy periods. Similarly, tracking cooling system data over time may reveal a gradual decline in performance. That early clue enables teams to resolve the issue before it escalates into a full-blown failure that impacts critical servers.

Once this data is collected, it's processed by advanced algorithms that utilize machine learning to identify patterns and risks. These tools identify any anomalies that fall outside the normal range and then forecast what might happen next. If certain types of hard drives or server components tend to fail after a specific number of operating hours, the system will flag them in advance. That allows staff to replace them before an outage occurs. This kind of insight helps avoid surprises and keeps the systems up and running without unplanned interruptions.

Network traffic is another area where predictive analytics shines. By monitoring slowdowns or heavy usage, the system can alert teams before network congestion occurs, preventing problems. It can also recommend changes to how resources are allocated across the network, ensuring a balanced and efficient performance even during peak usage times.

A major benefit of predictive analytics lies in its ability to improve maintenance. Rather than following a fixed schedule that may or may not align with actual wear and tear, teams can use data to determine the optimal time for repairs. This reduces unnecessary service work and ensures that the most critical parts are addressed first. For example, if cooling fans start to vibrate more than usual, that's a sign that something might go wrong soon. Predictive analytics can catch that change and alert the team, who can then replace the fan before it fails. It's the same for hard drives—tracking health data lets the team swap out aging drives before data is lost.

Looking beyond repairs, predictive analytics is also a powerful tool for capacity planning. It helps teams understand how usage is changing and what resources will be required in the future. Analyzing trends in power usage, server loads, and storage demand helps avoid both overbuying and running out of capacity. For example, if server demand is increasing gradually over the year, the data will indicate when upgrades should occur and what type of expansion is required. That way, decisions are based on facts rather than guesses. It also ensures that equipment arrives on time, avoiding delays that could impact business performance.

Planning for future storage needs works the same way. If storage usage is rising steadily, the system will highlight when more capacity is needed. That gives time to order new equipment, complete installations, and avoid last-minute scrambles. This kind of visibility helps data centers stay one step ahead of demand, ensuring the infrastructure can keep pace with business growth without incurring unnecessary expenses on upgrades.

Predictive analytics makes the data center smarter and more responsive. Instead of waiting for something to go wrong, teams can make informed choices that keep everything running smoothly. With fewer emergencies, better planning, and improved efficiency, data center operations become more stable, more affordable, and better prepared for the future.

Another important application of predictive analytics is reducing energy consumption and enhancing efficiency. Data centers consume a substantial amount of power, and finding more efficient ways to manage that power is crucial to lowering costs and minimizing their environmental footprint. With predictive analytics, data center teams can analyze energy use patterns, identify areas where energy is being wasted, and implement changes that conserve power without compromising performance. For instance, by examining the relationship between energy usage and factors such as room temperature and humidity, managers can optimize cooling systems to use less power while maintaining stable conditions. Similarly, reviewing server activity data can help implement more effective energy-saving methods, such as dynamic power scaling, which reduces energy consumption during periods of lower activity.

Making predictive analytics work well takes the right tools and skills. It begins with robust data collection systems and a solid infrastructure to manage the information. Software for analysis and clear data visualization is also a must. In many cases, a skilled team—such as data engineers or analysts—is required to manage the system and ensure the predictions are accurate and useful. This team also works on integrating analytics tools with the rest of the data center systems, which is crucial for maintaining smooth operation and ensuring the right data is available for analysis.

A successful rollout of predictive analytics goes beyond just installing software and collecting data. It also requires a shift in how the team thinks and works. The operations crew needs to shift from reacting to problems as they arise to planning and fixing issues before they occur. That kind of change takes time, but it also requires training. Staff must understand how predictive analytics works and how to use the results to guide better decisions in their daily tasks. Without buy-in from the team, even the best analytics system won't deliver real value.

It's also important to recognize that predictive analytics has limits. The accuracy of the system depends directly on the quality of the data it receives. If the data is incomplete, old, or wrong, the predictions won't be reliable. That's why it's essential to regularly check and fine-tune the models. As new data becomes available, teams should utilize it to enhance the system, ensuring that the insights remain relevant and useful. It's also worth remembering that no system can account for everything. Unexpected problems—such as power surges, extreme weather, or sudden equipment failure—can still disrupt things. Therefore, teams should always combine predictive insights with human judgment. The best results are achieved by combining data and experience.

The benefits of incorporating predictive analytics into a data center are evident. It helps cut costs by identifying problems early, optimizing resource utilization, and preventing failures before they occur. It keeps services reliable by supporting better maintenance planning and smarter scaling decisions. It improves efficiency by reducing wasted energy and enabling the team to work more efficiently and effectively. Just as important, it provides managers with the facts they need to make informed, strong decisions. This leads to better resource planning, fewer unexpected outages, and stronger overall performance.

By using predictive analytics, data centers can transition from constantly reacting to problems to proactively managing them before they arise. This shift helps increase uptime, lower costs, and make the whole operation more stable and efficient. Instead of guessing, managers can plan with confidence, knowing they have real data to back up their choices. With the right tools in place, predictive analytics can enhance the way data centers operate for the better.

Looking ahead, data will continue to play a larger role in decision-making within the data center. The ongoing improvements in predictive tools mean even more opportunities to boost efficiency and improve system reliability. As these technologies get smarter and easier to use, they will become a standard part of every well-run data center. Staying current with these changes will help teams stay competitive and fully prepared for whatever comes next.

Conducting a Post-Transition Assessment

The full handover of a data center from build to full operation isn't just a final step. It's the result of careful planning, clear communication, and consistent follow-through. What comes next is just as important—an in-depth review of how well the transition worked. This post-transition assessment isn't a quick once-over. It's a focused and structured effort to identify areas for improvement, highlight successes, and inform future enhancements. A clear strategy that utilizes proven methods and delivers measurable results helps the team gain a comprehensive understanding of its objectives. This isn't just about ticking boxes. It's about digging into the process to understand what worked, what didn't, and where improvements can be made for the long term.

The best way to run this assessment starts long before the move even begins. Key Performance Indicators (KPIs) and Critical Performance Indicators (CPIs) should be established during the planning phase. These KPIs and CPIs should match the goals of the data center project and reflect what the team defines as success. By keeping them clear and measurable, the team can track whether the results meet expectations. These metrics give you more than just numbers—they help tell the story of how the transition played out and where to focus next.

Uptime is one of the most important Key Performance Indicators (KPIs) and Critical Performance Indicators (CPIs). It shows how often critical systems remain online and run without issues. If the systems remain up nearly all the time, it means the transition worked well. This includes power, cooling, network, and other vital systems. Teams should use both detailed logs and automated tools to track this KPI and CPI. If uptime dips below the target, the team should investigate the cause and address it promptly.

Mean Time to Repair (MTTR) is another vital metric. It tracks how quickly the team can resolve issues when they occur. A short repair time means your team knows what to do, how to do it, and where to focus their efforts. It also demonstrates that the support systems are robust and the troubleshooting methods are effective. Keeping an eye on MTTR across all systems helps spot areas that may need more training, better tools, or clearer processes.

Power Usage Effectiveness (PUE), as mentioned earlier, helps the team assess how efficiently the data center utilizes energy. The number compares total energy use to the amount of energy used by actual IT equipment. A lower PUE means less energy waste and lower costs. This KPI and CPI should be tracked both during the build and after the move. If the number worsens after the transition, it's a red flag. The team should inspect airflow, cooling, server setup, and other areas to determine the cause of the drop. Capacity Utilization measures how well the data center uses its available space, power, bandwidth, and hardware. If the numbers indicate high usage, it means the planning accurately reflected the real needs. However, if usage remains low, it may indicate that the setup has more capacity than needed, resulting in wasted space and money. This metric helps guide future upgrades and changes to ensure that they avoid overbuilding or falling short of expectations.

Environmental metrics matter too. These include temperature, airflow, and humidity levels. They help ensure the environment is right for the equipment to work well. Tracking these numbers gives the team an early warning if something goes wrong. If the temperature or humidity shifts outside the safe range, it could indicate issues with the cooling system or air movement. Teams should investigate these changes immediately to prevent more significant problems down the line.

Ticket resolution time plays a major role in how well the operations team manages support and technical issues. This key metric measures the speed at which staff members close out tickets submitted by users or clients. A short resolution time often indicates that the support process is effective, the staff are skilled, and that troubleshooting steps are clear and concise. By examining this metric across different types of tickets, teams can identify which problems require the most time and where improvements might be most beneficial. It also helps determine if the helpdesk or technical team requires additional support, improved tools, or updated training. If response times start to grow, it's often a signal that something needs to be adjusted within the daily workflow or team structure.

Service Level Agreements (SLAs) serve as promises made to clients or internal users. These agreements outline the timeframe for the team to respond, fix, or maintain specific services. Keeping track of SLA performance helps confirm that those commitments are being met. If something falls short, it's essential to investigate the reason and take prompt action. Missing SLA targets too often leads to unhappy customers and can damage trust. Teams should not only track whether SLAs are met—they should also investigate the trends and causes behind any missed targets. This creates a fair and helpful picture of what's working and what needs attention.

To go beyond just the numbers, teams need to gather feedback from people who experienced the transition firsthand. This might include data center operators, IT staff, and in some cases, tenants or clients. Their input helps paint a more complete picture of the handover. Listening to feedback through interviews, surveys, or team debriefs adds context that raw data can't always show. Someone might flag an issue that didn't show up in logs but still affected performance or workflows. These insights often help identify blind spots or areas where written plans didn't match real-world experience.

It's essential to establish a clear plan for collecting data. That means knowing when and how to collect it, who's in charge of each part, and which systems will be involved. Effective monitoring tools and automated logs make this process easier. Teams should collect information from multiple sources, including ticketing systems, power logs, environmental monitors, and network tools, to gain a comprehensive view. Cross-checking information helps avoid blind spots and gives better insights. The more accurate and complete the data, the more useful the final analysis will be.

Once all the data is collected, the team needs to analyze it carefully. That means comparing results to what was originally expected. If something misses the mark, the team should look at why. Root cause analysis helps identify the underlying reason behind problems, rather than just addressing the surface-level issue. The analysis should examine both what went wrong and what might cause problems in the future. Some issues don't appear immediately, but small warning signs in the data can help identify them early if people know where to look.

Everything learned from the assessment should be included in a report that's clear, direct, and easy to use. The report should include all the collected data, its interpretation, what went well, and what needs improvement. It should also give action steps and improvement plans based on the findings. Every stakeholder—whether they're from operations, IT, safety, or vendor teams—should receive the report. That way, everyone has the same understanding and can follow through on the action steps together. A strong report gives direction, promotes accountability, and keeps teams focused on shared goals.

The transition review is not just something you do once; it's a continuous process. It should be part of an ongoing routine. Teams need to continually monitor key metrics, check systems, and implement small improvements. This ongoing check-in ensures the data center operates smoothly and helps identify issues before they escalate. It also fosters a culture where the team remains alert and takes prompt action. Instead of reacting to problems, the team becomes good at preventing them. This mindset helps protect the entire facility and the information it stores. When done right, this cycle of review and adjustment builds a data center that stays strong, efficient, and reliable well into the future. It also helps ensure that the money, time, and effort spent on the build continue to deliver long-term value.

Identifying Areas for Improvement and Optimization

The post-transition assessment described earlier offers more than just a collection of numbers or observations. It provides a comprehensive view of how the data center performs after the transition from construction to daily operations. However, the real power of this review doesn't come from merely identifying problems. It stems from utilizing the insights gained to enhance the facility's operations and improve ongoing performance. To make that happen, the team needs to follow a clear and focused process—one that goes beyond quick fixes and aims for long-term improvement and reliability.

Teams begin by taking a close look at the data they've collected.

This includes examining Key Performance Indicators (KPIs) and Critical Performance Indicators (CPIs) such as uptime, Mean Time to Repair (MTTR), Power Usage Effectiveness (PUE), capacity use, environmental conditions, ticket resolution times, and how well SLAs are being met. By comparing actual performance to expected benchmarks, it becomes easier to identify areas where things fall short. If any metric shows large gaps from its goal, the team needs to look deeper. For instance, if the MTTR for a certain piece of equipment is higher than expected, that might point to issues like missing spare parts, unclear instructions, or a lack of proper training. Alternatively, if the PUE numbers consistently exceed the planned values, the cooling system may be the weak point. It could require cleaning, retuning, or even upgrading to a more efficient solution.

The numbers alone don't always tell the full story. Direct feedback from individuals working in or with the data center provides valuable context. Input from staff, tenants, and IT teams helps explain why the numbers appear as they do. For example, the data might indicate that ticket resolution is slow, but the support staff might explain that the ticketing system is outdated, clunky, or difficult to use. This kind of first-hand experience helps teams understand the real problems and makes it easier to come up with useful solutions that actually work on the ground.

After reviewing the data and hearing from those involved, the team should take the next step—finding the root causes behind the issues. That means digging deeper, not just treating symptoms. Root Cause Analysis (RCA) facilitates this process by utilizing straightforward and practical tools, such as the "5 Whys" or cause-and-effect (Fishbone) diagrams. If certain server racks consistently show high heat levels, for example, the team could ask "why" repeatedly until they reach the source. They might discover poor airflow design or missing tiles beneath the floor that disrupt cooling.

Fixing that issue would involve adjusting the layout or adding extra airflow support.

Once the problems are fully understood, it's time to create real, focused action plans. These plans should be clear and practical— not just general ideas. Using SMART goals helps make this possible. A goal might be, "Cut the MTTR on equipment X by 20% over the next three months through focused training and spare part readiness." The plan needs to specify who is responsible for what, by when, and how progress will be monitored and evaluated. It might include updated procedures, training workshops, or software upgrades. Assigning owners to each step makes sure nothing gets lost along the way. Teams should also establish regular checkpoints to monitor progress and make adjustments as needed.

Implementing these strategies requires careful consideration and effective coordination. Teams need to assign the right resources, work closely with all involved departments, and establish a clear system to track every step taken. To stay on course, they should conduct regular reviews to assess whether things are working, identify any issues early, and adjust plans as needed. Open communication and honest updates keep everyone informed, fostering trust and teamwork. Holding weekly project meetings allows everyone to discuss their progress, clarify any confusion, and resolve problems together. Using visual tools like dashboards helps give a clear view of how close each team is to hitting their goals and where they might need support. Looking past the first round of improvements, building a solid system that supports ongoing progress plays a major role in keeping the data center running at its best.

That means weaving continuous monitoring and regular feedback into everyday tasks. Installing automated systems that track KPIs and CPIs in real-time helps identify trends and take action early. Gathering feedback through staff check-ins or short tenant surveys makes it easier to hear directly from those working in or using the space. Teams can then use this feedback to refine their procedures and prevent issues from escalating into serious problems.

Staff play a big role in helping the data center improve over time. Managers should create an environment where people feel safe to point out issues, share ideas, and participate in solving problems. Encouraging everyone to help find the cause of an issue and shape better ways of doing things leads to smarter and smoother operations. Ongoing training also benefits not only technical areas but also communication, problem-solving, and teamwork. Hosting regular training sessions provides people with the necessary tools to perform their jobs effectively and prepares them for future challenges. When major problems do occur, the team should meet to review what went wrong, what it affected, what they've learned, and how to prevent it from happening again. These post-incident reviews help everyone learn and continually raise the bar. The drive to improve can't sit on the sidelines—it needs to be at the heart of daily operations.

Keeping a close eye on KPIs and CPIs, listening to the staff, and hearing from tenants all work together to shape future changes. When teams go through this cycle—watch, analyze, improve, apply—they build a system that continually improves. This approach enables the data center to remain efficient and prepared to handle any future challenges that may arise. When facility managers view continuous improvement as part of their daily thinking, rather than just an additional task, they safeguard the long-term strength, safety, and value of their facility.

Tracking whether this approach works is just as important as implementing it. Clear, measurable goals—such as reducing energy waste, increasing system uptime, or decreasing repair time—help teams see progress and keep their efforts focused. By sharing regular updates on these results, teams can demonstrate real progress, explain why these changes are significant, and make a compelling case for continued support and funding. This steady check on results ensures the efforts remain sharp, timely, and useful. The point isn't just to fix what's broken—it's about always looking for ways to do things better. When teams adopt this kind of mindset, they help the data center become stronger, more secure, and a better long-term investment.

Developing a Continuous Improvement Plan

Developing a comprehensive, continuous improvement plan requires a structured approach, moving beyond a simple list of tasks to a strategic roadmap for sustained optimization. This plan should be a living document, regularly reviewed and updated based on ongoing performance data and feedback. The first step involves prioritizing the areas for improvement identified during the post-transition assessment.

This prioritization should be based on a combination of factors, including the severity of the impact on operations, the feasibility of implementation, and the potential return on investment. A simple matrix, ranking issues by impact and feasibility, can be a valuable tool in this process. High-impact, high-feasibility issues should be addressed first, while low-impact, low-feasibility issues may be deferred or re-evaluated later. Once prioritized, each area for improvement should be addressed with a specific action plan. This plan should outline the specific actions to be taken, identify the responsible parties, specify the required resources, and include a realistic timeline for completion. For instance, if the post-transition assessment revealed inefficiencies in the cooling system, the action plan might involve:

- Action: Conduct a comprehensive audit of the cooling system to identify bottlenecks and areas for improvement.
- Responsible Party: Facilities Engineering team, in collaboration with the vendor.
- Resources: Specialized testing equipment, engineering expertise, and potential contractor assistance.
- Timeline: Complete the audit within one month; implement recommended changes within three months.

Similarly, if staff training was identified as a deficiency, the action plan might involve:

- Action: Develop and deliver a comprehensive training program on critical systems and procedures.
- Responsible Party: Training department, in collaboration with subject matter experts from operations and IT.
- Resources: Training materials, dedicated training facilities, instructor time.
- Timeline: Develop the training program within two to four weeks; deliver training to all relevant staff within one to three months.

For each action plan, measurable Key Performance Indicators (KPIs) and Critical Performance Indicators (CPIs) should be defined to track progress and evaluate success. These CPIs should be specific, measurable, achievable, relevant, and time-bound (SMART). For example, for the cooling system audit, the Key Performance Indicator (KPI) and Critical Performance Indicator (CPI) may result in a 5% reduction in PUE within six months of implementing the recommended changes. For the staff training program, the KPI and CPI might be a reduction in MTTR for specific system failures by 15% within three months of training completion.

The continuous improvement plan should also outline a system for monitoring progress and providing regular updates. This may involve submitting weekly or monthly progress reports, utilizing dashboards that visualize key KPIs and CPIs, and holding regular meetings to review progress and address any challenges that arise. Transparency is crucial; all stakeholders should have access to the plan and progress reports, ensuring buy-in and collaborative problem-solving. The use of project management software can significantly aid in tracking progress, managing tasks, and facilitating effective communication across teams. SharePoint or similar platforms can serve as central repositories for the plan, progress reports, and related documentation.

The plan should also incorporate mechanisms for ongoing feedback and evaluation. Regular surveys of staff and tenants can provide valuable insights into areas for improvement that may not be readily apparent through quantitative data analysis. These surveys should be designed to elicit specific feedback on operational processes, system performance, and overall satisfaction. Furthermore, a system for reporting and addressing incidents should be in place, ensuring that all issues are documented, analyzed using root cause analysis methodologies, and addressed promptly. This ensures that even minor issues do not escalate into major problems.

A critical component of the continuous improvement plan is establishing a culture of continuous learning. This involves empowering staff to identify and report problems, participate in root cause analysis, and contribute to the development of improved procedures and processes. Regular training programs, both technical and soft skills focused, are essential. Regular post-incident reviews, where the team analyzes the cause and impact of significant incidents, identifies learning points, and develops preventative measures, are also vital for continuous learning. Encouraging a culture of open communication and feedback is paramount to success.

The continuous improvement plan should be integrated into the overall data center operations manual, making it a core component of the facility's operational philosophy. This ensures that continuous improvement is not viewed as a separate project, but rather as an ongoing process embedded within daily operations. Regular reviews of Key Performance Indicators (KPIs) and Critical Performance Indicators (CPIs), coupled with ongoing feedback, provide the necessary insights to inform ongoing optimization efforts. This creates an iterative cycle of monitoring, analysis, improvement, and implementation.

The plan should also include contingency plans to address unforeseen challenges or setbacks. Unexpected events, such as equipment failures or external disruptions, may necessitate adjustments to the plan.

Having pre-defined escalation procedures and alternative solutions can help mitigate the impact of such events. Regular review and updating of the plan ensure it remains relevant and adaptable to changing circumstances.

Finally, the success of the continuous improvement plan must be measured. This involves setting measurable targets for improvement in specific areas and tracking progress against those targets. Regular reporting on the achieved improvements showcases the tangible value of the program, justifying the investment and highlighting the importance of this ongoing endeavor.

This consistent evaluation ensures that the continuous improvement process remains focused, relevant, and impactful, ultimately contributing to the long-term success, reliability, and cost-effectiveness of the data center. It's about creating a self-improving system, constantly striving for optimization and resilience, ensuring the data center remains a top-performing asset for years to come. The continuous improvement plan isn't just a document; it's a commitment to operational excellence.

Monitoring Performance and Tracking KPIs and CPIs

Keeping a close eye on how the data center performs after the transition isn't something you do just once—it's a constant responsibility. The real success of the handover doesn't stop with the switch from construction to live operations. It depends on how well the facility runs day after day. This ongoing success relies on building a strong system to monitor performance and track Key Performance Indicators (KPIs) and Critical Performance Indicators (CPIs). These KPIs and CPIs act as guideposts. They provide clear insights into the health of the systems and the overall effectiveness of the entire system. They also help teams identify early warning signs, allowing them to intervene before small problems escalate into major disruptions. In the long run, this reduces downtime and helps maintain smooth operation.

Choosing the right KPIs and CPIs makes all the difference. A good mix provides a comprehensive view of the data center's performance, encompassing key areas such as power usage, system maintenance, failure rates, temperature, humidity, space utilization, and operational costs. Each KPI and CPI should align with a specific goal and be easy to track and understand. If a metric doesn't directly help improve performance or identify issues early, it shouldn't be included. Well-defined KPIs and CPIs save time, focus attention, and help teams take the right actions more quickly.

Power usage effectiveness, or PUE, is one of the most important metrics. It shows how much of the total energy the data center uses compared to the energy that actually powers the IT equipment. A lower number means less waste and better efficiency. Tracking this over time helps teams catch problems early. If energy use starts creeping up, the team can examine the cooling system, review settings on the IT equipment, or determine if it's time to upgrade to more efficient equipment. Spotting these trends allows teams to make changes before energy bills increase or the system slows down. Even a slight improvement in PUE can result in substantial savings and improved performance.

Mean Time To Repair (MTTR) refers to the average time required to resolve an issue once it has occurred. A low MTTR means the team identifies the issue quickly and restores service without delay. Watching this number helps identify slow parts in the process—perhaps the right parts aren't available, staff need better training, or the repair steps are too complicated. By identifying and resolving these trouble spots, the team reduces repair times and keeps systems online more frequently. For example, if fixing one type of server always takes too long, it may be time to rethink how the team handles those repairs or provide them with more support and tools to work more efficiently.

Mean Time Between Failures (MTBF) looks at how often things break. A higher MTBF means systems operate reliably for longer periods without issues. This indicates that the gear is well-maintained and the team is performing a good job. If that number starts dropping, something might be wearing out, or something in the process needs attention. Teams can take that as a sign to check equipment more frequently, replace aging hardware, or fine-tune their management of those systems. Comparing MTBF data with other numbers—like temperature or humidity levels—can also reveal hidden issues. For instance, if one room experiences more breakdowns and also runs hotter than the others, it's a strong indication that the environment needs adjustment.

Environmental monitoring plays a crucial role in maintaining the optimal operating conditions of the data center. The team must maintain a constant vigilance over temperature and humidity to prevent issues such as hardware damage, performance degradation, or complete system failure. When these conditions move outside the safe range, the risk of disruption increases.

Reliable monitoring tools that provide real-time updates enable staff to identify issues promptly and address them before they cause significant damage. The faster they respond, the better they can protect the equipment and maintain stable operations. These alerts save time and money by helping prevent bigger issues down the line.

Looking back at past environmental data also helps the team understand how different seasons or patterns may affect the center. For example, if a certain time of year always brings higher temperatures, they can plan by adjusting cooling systems in advance. This kind of proactive thinking makes the facility stronger and reduces the number of surprises they face. Understanding these patterns also helps with planning long-term upgrades and avoiding unexpected repair costs.

Capacity utilization provides valuable insights into how effectively the space and equipment are being utilized. Tracking how much of the servers, storage, and network bandwidth is in use helps guide better planning. If usage is too high, the center might need to expand. If it's low, it may be time to consider consolidating or adjusting resources. Regularly monitoring capacity ensures that the center can handle growth without encountering slowdowns. This type of insight helps avoid performance problems and supports smarter decisions about future needs. It's also important to match this data with business goals, so that expansion lines up with expected demand.

Operational costs reveal the true cost of maintaining the data center. This includes everything from power and cooling to staff and routine maintenance. Keeping track of these costs helps keep spending under control while still maintaining solid performance. By analyzing the data, the team can identify opportunities to save money, such as reducing power consumption, optimizing workflows, or streamlining specific tasks. If something starts to cost more than planned, the data makes it easier to catch and fix the problem. Comparing actual expenses against the budget also keeps the financial side on track and supports better decisions in the long run.

A good monitoring system is key to making all of this work. It should pull in data from various parts of the data center, including power usage, temperature, and humidity sensors, as well as IT system logs, and display all of this information in one place. Staff should be able to view real-time data in a simple dashboard and receive alerts when an issue arises. These alerts must be concise and clear so that the team can respond promptly and effectively. Tools like graphs and charts help turn data into useful insights, making it easier to spot trends and make smart decisions based on what the numbers are saying.

The system itself must work seamlessly with existing tools and infrastructure to provide a comprehensive picture. It should store data securely and keep older records available for later review. Looking back at this history helps spot long-term trends and compare current performance to past results. The system should be reviewed regularly to ensure it's functioning properly. These checkups help identify and address any issues, and highlight areas where improvements could make it even better.

No monitoring plan works without the right people in place. Staff need to understand what the data means and what to do when it deviates from the expected track. Regular training helps them stay sharp and know how to act quickly when problems arise. Everyone should know their role in a crisis and how to report or escalate a problem. Good communication keeps the team working as a cohesive unit, which makes the center safer and more reliable overall.

All data collected through this type of tracking should be connected to the existing improvement plans. Teams should discuss these numbers in their regular meetings and use them to identify areas that require more attention.

Key Performance Indicators (KPIs) and Critical Performance Indicators (CPIs) should be reviewed periodically to ensure they remain aligned with the operational goals. When the business changes, the metrics may need to change as well. Keeping this process active ensures the center continues to improve and operates at its best.

Checking on the success of the monitoring plan is just as important as setting it up. The team should compare performance numbers to the goals they set, and these comparisons should be part of regular reports.

These updates help demonstrate whether the time and money spent on monitoring are yielding results. If the results are good, it proves the value of careful tracking and quick response. The effort to continually improve and monitor performance isn't just another task—it's a smart way to protect the data center and keep it running strong for the long run.

Adapting to Future Challenges and Trends

The successful transition of a data center from construction to full operation marks just the beginning of a much longer journey. The data center industry is dynamic, characterized by constant innovation, evolving regulations, and shifting business needs. To keep the facility performing well, a critical facilities manager needs to take a forward-thinking approach. That means staying sharp, learning continuously, planning with intention, and remaining open to change as it arises. Standing still leads to falling behind.

Technology continues to shape the future of data centers more than almost anything else. As computing power, networks, and storage systems evolve rapidly, the data center must keep pace. Rigid designs built for specific hardware no longer last very long. Smart managers build flexibility into the design from the start. Modular layouts let them swap in new equipment or scale up when needed without shutting everything down.

As new systems require more power and generate more heat, it's not enough to just add new racks. Planning must encompass improved cooling, optimized cable layouts, and enhanced network paths to ensure optimal performance. If the team doesn't prepare for this ahead of time, they could face major delays, high costs, or wasted resources.

The move toward virtualization and cloud services adds even more complexity to managing a data center. Virtual environments enable teams to get more out of the same hardware, but they also make everything more difficult to track and manage. Managers need to understand how virtual systems impact factors such as energy use and airflow, and they must adjust the systems accordingly to optimize their performance. Keeping up with this shift means staff need better training in server management and networking. Teams must also take a closer look at their security plans. Virtual machines open the door to new kinds of threats, and the center needs solid systems in place to keep data safe. Firewalls, detection tools, and fast-response plans are no longer optional; they are essential. They're part of doing the job right and keeping customers safe.

Government rules regarding power use, privacy, and the environment are constantly evolving, and this trend is not expected to slow down. Facilities that fail to stay up to date run the risk of violating laws and incurring penalties. Adhering to these rules requires time and effort, but it's worth it. Staff must be aware of the new regulations and understand how to comply with them. For example, new laws may require changes to cooling setups or upgrades to make systems more energy-efficient. Privacy rules, such as GDPR or CCPA, also require robust policies on who can access data and how it's stored.

Failing to address these risks can result in legal issues, substantial fines, and reputational damage to the company. Getting compliance right protects the business and builds trust with clients and partners.

The way a business grows or shifts direction also plays a big role in how the data center must be run. A company might grow faster than expected, merge with another company, or change its service delivery approach. Each of these changes puts new demands on the data center's systems and layout. A flexible setup enables the team to adjust quickly, without wasting time or resources. Forecasting tools can provide a clear picture of what's to come, allowing the team to plan resources accordingly. By analyzing patterns in data usage and storage, managers can plan for new server needs or extra bandwidth. This kind of planning helps avoid traffic jams, delays, or performance issues later on. It also provides better service for users and reduces surprises.

The ongoing learning and professional development of the data center team play a vital role in staying ahead of constant changes. Critical facilities managers must put staff development at the top of their list. Training programs should focus on practical skills, real-world knowledge, and hands-on experience that enable the team to handle new technology and challenges with confidence. This includes instruction on current tools, updated security procedures, and smart ways to manage systems efficiently. Certifications and regular development courses help build stronger teams with the knowledge and expertise to support the center as it grows and evolves. A well-trained staff is the backbone of a reliable operation.

Managers should also make research and development a regular part of operations. Keeping up with new ideas takes curiosity, testing, and a willingness to try things that haven't been done before. Research may involve testing new cooling systems, exploring energy-saving equipment, or examining how artificial intelligence can enhance the center's operations. These ideas may not always work out, but they propel the operation forward and give it an edge. Attending trade events, subscribing to respected industry journals, and connecting with peers in the field also help staff stay up to speed and bring back ideas worth trying.

Strong teams thrive in a culture where people feel free to suggest new ideas and speak honestly about what works and what doesn't. Leaders can foster this kind of culture by soliciting feedback, actively listening to it, and demonstrating that it leads to tangible action. They should regularly review performance metrics such as PUE, MTTR, and capacity utilization to inform better decisions. Teams should treat those numbers not as reports to file away but as tools to solve problems and reduce waste. Metrics are more than statistics—they guide better choices, improve results, and help spot small issues before they become big ones. A strong risk management plan must be in place to protect both the data center's equipment and its day-to-day operations. Managers need to consider the full picture—identifying weak spots, assessing their likelihood of causing trouble, and evaluating the potential harm they could inflict on the operation.

Once risks are clear, teams can develop solid plans to mitigate the impact of any potential issues. This includes setting up business continuity and recovery plans that keep things running during power outages, fires, cyberattacks, or major hardware issues. It's about planning for the worst so the team can stay calm and respond quickly when trouble hits.

Good communication holds everything together. Clear and open communication between team members, across departments, and with outside vendors and tenants facilitates the avoidance of confusion and enables the prompt handling of issues as they arise. Communication should never be left to chance. Managers should establish clear systems with agreed-upon channels and rules for sharing updates. Transparency fosters trust, and people who trust each other work more effectively together. When teams report their progress, problems, and upcoming plans to leadership, it keeps everyone on the same page and provides leaders with the information they need to make informed decisions.

It takes more than just solid equipment to run a successful data center. Teams must build a habit of learning, testing, improving, and communicating. They need to understand the tech, follow the rules, adjust to the business, and plan for the unexpected. When the team stays ready, the operation stays strong. Adapting to change means incorporating good habits into your daily routine and remaining flexible, regardless of what the future holds. Critical facilities managers who live by these principles lead teams that not only survive but thrive, even as demands shift and the industry changes shape. The work is never done—but that's what makes success possible.

Acknowledgments

This book would not have been possible without the contributions of numerous individuals. First and foremost, I want to thank Mohammed Talha, Commissioning Engineer at JLL with a BEng in Mechanical Engineering and an MSc in Information Technology, whose expertise provided invaluable insights, especially with the Direct liquid and Immersion cooling section. I also extend my deepest gratitude to the critical facilities managers, chiefs, operations engineers, subject matter experts, project managers, general contractors, and IT professionals who generously shared their expertise and experiences, providing invaluable insights and real-world examples that enrich this guide.

Their willingness to share their knowledge and best practices has been instrumental in shaping the content and ensuring its practical relevance.

Appendix

Appendix A: Sample Data Center Transition Plan Template (Includes sections for timelines, responsibilities, risk assessments, and communication protocols).

Appendix B: Checklist for Pre-Operational System Testing (Covers power, cooling, security, network, and environmental monitoring systems).

Appendix C: Example SLA Agreement between Data Center and Tenant.

Appendix D: Template for Cabinet Board Approval Request.

Appendix E: Glossary of Acronyms and Terms (This list supplements the main glossary, providing additional technical terminology and abbreviations relevant to specific examples discussed in the book).

Glossary

- **Availability:** The percentage of time a system or component is operational and accessible.
- **Business Continuity Plan (BCP):** A plan outlining how an organization will continue operations during and after a disruptive event.
- **Commissioning:** The process of verifying that systems and equipment are installed and functioning correctly.
- **Data Center Critical Facilities Manager (CFM):** An individual responsible for the operation and maintenance of critical infrastructure, including data centers.
- **Data Center Infrastructure Management (DCIM):** Software used to monitor and manage data center resources.
- **Disaster Recovery Plan (DRP):** A detailed plan to recover data and systems after a disaster.
- **Mean Time Between Failures (MTBF): The** average time between failures of a system or component.
- **Mean Time To Repair (MTTR):** The average time required to repair a failed system or component.
- **Power Usage Effectiveness (PUE):** A metric that measures the efficiency of a data center's power usage.
- **Service Level Agreement (SLA):** A contract defining the performance expectations between a service provider and a customer.

References

ASHRAE TC 9.9 *Standard for Data Centers: Environmental Guidelines for Data Processing Environments,* accessed from https://www.ashrae.org/file%20library/technical%20resources/bookstore/ashrae_tc0909_power_white_paper_22_june_2016_revised.pdf

Uptime Institute Data Center Tier Standard: *Topology,* accessed from https://uptimeinstitute.com/resources/asset/tier-standardtopology

Author Biography

Yekini K. Tidjani is an experienced Data Center Critical Facilities Manager with a decade of experience in the data center industry. His background includes extensive experience in data center construction, operations, management, control systems, and IT encompassing all aspects from initial design and build to ongoing maintenance and optimization. He has a proven track record of successfully transitioning data centers from construction to full operational status, minimizing disruptions and maximizing efficiency.

Yekini K. Tidjani holds a Bachelor of Science Degree in Electrical Engineering and Communications Technology, as well as an Associate of Applied Science Degree in Computer & Electronics Engineering Technology. His expertise encompasses operations management, construction oversight, controls, electrical and mechanical cooling systems, risk management, operational efficiency, and customer support.

He is passionate about sharing their knowledge and experience to help others succeed in the demanding field of critical facilities management. He is a husband to a wonderful wife, to whom he has been married for nearly two decades, and they have four children.

Yekini K. Tidjani is also a tennis player and enjoys reading theological books. He is from West Africa, specifically Togo, and has been living in Houston since 2000.

Made in the USA
Monee, IL
04 November 2026

67384097R00125